GLOBAL BEAUTY, LOCAL BODIES

Also edited by Erynn Masi de Casanova and Afshan Jafar:
Bodies without Borders

GLOBAL BEAUTY, LOCAL BODIES

Edited by

Afshan Jafar
and
Erynn Masi de Casanova

First published in 2013 by
PALGRAVE MACMILLAN®
in the United States—a division of St. Martin's Press LLC,
175 Fifth Avenue, New York, NY 10010.

Where this book is distributed in the UK, Europe and the rest of the world,
this is by Palgrave Macmillan, a division of Macmillan Publishers Limited,
registered in England, company number 785998, of Houndmills,
Basingstoke, Hampshire RG21 6XS.

Palgrave Macmillan is the global academic imprint of the above companies
and has companies and representatives throughout the world.

Palgrave® and Macmillan® are registered trademarks in the United States,
the United Kingdom, Europe and other countries.

ISBN: 978–1–137–37866–8

Library of Congress Cataloging-in-Publication Data is available from the
Library of Congress.

A catalogue record of the book is available from the British Library.

Design by Newgen Knowledge Works (P) Ltd., Chennai, India.

First edition: December 2013

10 9 8 7 6 5 4 3 2 1

For Aleena and Lilah, who make my life beautiful
A. J.

A Soledad, la fuerte y bella
E. M. C.

Contents

ILLUSTRATIONS

Bodies, Beauty, and Location: An Introduction

Afshan Jafar and Erynn Masi de Casanova

Warning: A quick glance at the title of this book may lead to a dizzying constellation of questions. When talking about beauty, which manages to be simultaneously abstract and ephemeral, embodied and concrete, what do the dimensions of global, transnational, national, and local mean? Since these very geographic terms and categories are contested, what can we really learn about beauty by using them? What's new about our approach in this volume? Is the globalization of beauty and body practices simply a diffusion of Western ideals, as is commonly believed, or do individuals and societies negotiate and resist Western norms? How does the interaction of local and global ideas about bodies produce particular forms of embodiment? To address these questions, this innovative volume debuts original research and personal reflections on: media depictions of Nordic metrosexual athletes; prostitutes working at the US/ Mexico border; beauty ideals among Somali migrants to Kenya; and the popularity of nose jobs among Iranian women, among other timely and understudied topics.

In this introduction, we take the processes of globalization, and the embeddedness of bodies in transnational relations, as a starting point, rather than simply looking at how beauty is understood in a series of locales around the world. The authors of the chapters in this volume have done a wonderful job of covering the key texts and theories, so this introduction does not provide an exhaustive review of the literature on beauty and the body. Exhaust*ive* can often become exhaust*ing*, and we don't want readers to skip the introduction! The chapters in this volume include scholarly essays based on empirical research as well as shorter personal reflection pieces in which our authors discuss their own experiences with globalization and beauty. In this introductory chapter, we alternate between our personal experiences of globalization and beauty, posing questions about our subject and highlighting some of the themes that tie the book's chapters together.

*   *   *

FROM SNOW BLACK TO GORI MADAM

I couldn't have been more than five years old when I first realized the significance of skin color. My *khala* (mother's sister) took it upon herself to impart what she considered to be a valuable life lesson.

"Now listen to me, Afshan," she said. "You realize that you are not *gori* [light-skinned; but also a term for white people] like your sister is. Everybody will always like her, and talk about how beautiful she is and will always give her more attention because she is fair [skinned]. But if you work hard at developing a nice personality and are kind and respectful, then people will like you too."

With those devastating words she was done with her life lesson and moved on. I ran to my mother in tears, devastated that I was somehow less likable, less beautiful, and less wanted simply because I had darker skin than my sister. My mother told me a "secret" that calmed me down for the moment. She told me that God makes his favorite people darker—ostensibly to be able to spot them from afar, I'm not exactly sure. My five-year-old mind did not want to question the logic of an argument that reassured me being darker was a sign of being special.

My mother's message only soothed me for a while. As I grew up she couldn't shield me from the message that surrounded me: My darker skin color was not desirable. Another moment that sticks out above the jumble of memories from childhood is when I was given the role of the wicked Queen in *Snow White* in my elementary school play in fifth grade. When I declared excitedly to a cousin of mine that I was going to act in *Snow White*, he laughed and said, "Are they going to re-name the play *Snow Black*?" Strangely, I am not even considered *very* dark by Pakistani standards. But I am certainly not "fair."

Then, there were the countless home "remedies"—as if we were afflicted with a disease—which we all used unthinkingly, to make ourselves "fairer." Rule of thumb for home concoctions: if it looks white, it will make you light(er), if it looks dark, it will make you dark(er). So we often dipped cotton wool in milk and rubbed it all over our faces, not only because it was a good toner/conditioner for the skin, but in the hopes that it would take away our darkness. Or you would hear somebody reprimand a young girl that she shouldn't drink tea because it will make her dark, like tea. Of course the antidote here, again, was milk!

Even some Pakistani wedding traditions revolve around beautifying (lightening) a bride's skin tone. One event, known as the *Mayun*, takes place days before the wedding and involves the application of *ubtan* (a facial/body mask of various ingredients including turmeric, which is said to lighten the skin) to the bride's face and body. On the wedding day itself, it is common to see brides wearing a much lighter shade of foundation than their actual skin color, in order to appear more "beautiful."

When I was 18 years old, I left Pakistan for the first time and came to the United States for college. Suddenly, my skin color, which had been a mark of my inferiority, was now "exotic." But even as an 18-year-old, the joy of being labeled "exotic" was short-lived. At the time I couldn't voice my discomfort with the label "exotic"; all I knew was that it bothered me. Over the years, I have come to realize what it was: exotic implies difference—a deviation from the "normal," the "standard." Being labeled exotic was similar to giving that label to an animal or a bird at the zoo. You can marvel at them in the zoo, but you wouldn't want them in your house.

And so it was. I could not escape the relevance of the color of my skin no matter where I went. In Pakistan it was something that needed to be remedied, needed to be counterbalanced by other things I could offer that might set me apart from being ordinary. But in the United States, it was what marked me as different, as the unknown, mysterious, Other. In my time here, I have been mistaken for Mexican, Puerto Rican, Colombian, Arab, and Indian—Cherokee Indian.

A few years ago when I was at the Islamabad airport, in Pakistan, something very strange happened. An airport worker, while talking to my brother, referred to me as the "*gori* madam" (light-skinned lady/white lady). Where was that five-year-old girl whose aunt had told her about the disadvantage of her dark skin? How and when did she get replaced by a "*gori* madam?" Sure, after living in the United States I had lost some of the perpetual dark tan that I had growing up in the Pakistani sun. But I was still dark-skinned. Was it that the term *gori* implied something more than skin color in this context? At the time, I had a very short, spiky haircut, was wearing jeans and a T-shirt and was on my way back to the United States alone. I believe all of that signaled a particular privileged status about me and my location in the Western world. And privilege, especially associated with the West, could not be defined as dark. Privilege was *seen* as the right of a *gori* madam, so I became a *gori* madam for the first time in my life.

Alas, after all these years, I had come to realize that my skin color was not an inherent attribute. How my skin is perceived has as much to do with people's notions of privilege and status (and whether I have it or not), as it does with the particular shade of brown my skin happens to be.

If only my aunt had told me this instead when I was five years old. It would have been a far more interesting life lesson.

—Afshan Jafar

*   *   *

Progress, Modernity, and the Other

There have been changes more recently in the reception of dark-skinned models and actresses in South Asia. For instance, we now see many more dark-skinned models and actresses in magazines and playing the lead roles in movies and television dramas. It is no longer required to be light-skinned to be a model. However, at the same time when we are seeing these niches develop for dark-skinned women, it is interesting to note that their incorporation happens in limited ways. For instance, they are still not defined as beautiful or pretty but rather as "hot" or "attractive." It seems to be a self-exoticization of dark skin, where it is defined as a deviation from the standard of light-skinned beauty. This theme is of course echoed in the Western context as well as seen in the personal reflection above, and also in the scholarship on race and beauty in the Western context, where classic beauty is defined as white and women of color are depicted as hot, sexy, or downright animalistic. So while the increasing global coverage of brown-skinned celebrities like Beyoncé, Halle Berry, Eva Longoria, and Jennifer Lopez, to name a few, may have provided an impetus to create niches for brown-skinned women in the media globally, we see that the "local" ideals of light-skinned beauty remain intact and are reinforced by the global hierarchy of skin color. In a strange twist, we see many brown-skinned women winning Miss World and Miss Universe pageants. But they do so only by winning their national pageants first, which often means going through a strict regimen of skin lightening and skin bleaching. This belies the false assumption that people easily fall into: that skin lightening is an attempt to mimic a Western and Caucasian ideal of beauty. On the contrary, fairness is often seen as a deeply entrenched, local ideal of beauty. India and South Asia more generally provide an excellent example. Scholars are unsure about the origin of fairness as a beauty ideal in South Asia. Most agree that it predates the British colonization of the subcontinent, though that certainly reinforced and exacerbated colorism in the subcontinent (Vaid 2009).

One may ask why we chose to do a book that focuses entirely on beauty. Is this a significant enough aspect of our lives to merit such scholarly consideration? There are many scholars who have shown the relevance of beauty in people's lives (Hamermesh 2011; Rhode 2010). There are many who have talked about the white ideal of beauty and its impact on people of color (Hill 2002; Thompson and Keith 2001). There has also been some important scholarship on colorism (prejudice and discrimination based on skin color) and its impact on people's lives: including being able to secure good jobs, be promoted, and find a spouse (Hill 2000; Hunter 2002; Russell, Wilson, and Hall 1992; Vaid 2009).[1] So the relevance of one's appearance has been well established. However, a discussion of globalization and global power relations is often missing in these discussions.

Why is it, for instance, that while most Americans probably couldn't name the current Miss America, much less the last few, most people in India can rattle off a list of the last few Miss India winners, and take great pride in Indian women winning Miss World or Miss Universe? What does the political career of Irene Sáez—a former Miss Venezuela who won Miss Universe in 1981—as a mayor, governor, and presidential candidate, tell us about the pride with which pageant winners are embraced in other nations?

As the chapters in this volume show, bodies—whether they be pageant contestants, models, sex workers, or others—hold meaning beyond the individual: they are also symbolic of the progress or modernity of a nation. When a Nigerian or an Indian woman wins an international pageant, it is seen as an inversion of global power relations. As a contestant of the Miss India pageant put it, "Look, I come from a slummy country and here I am! I am going to wear my saris and win!" (Dewey 2008, 159).

The disciplining and modifying of bodies often has the implicit aim of embodying an idea of progress, development, and modernity that the world can relate to, and that nations measure themselves by. The idea of progress and modernity, often combined with the pseudo-scientific language of health, fitness, intervention, technology, and discipline, appeals to nations otherwise seen as "backward." These women's bodies, and increasingly men's bodies too, become one way in which nations claim a seat at the table of modernity and progress. As people modify their bodies, whether through skin lightening, working out, losing weight, or plastic surgery, they are seen as symbolic of a nation that is developing and evolving.

The theme of modernity is intertwined with the ideal of beauty, and its corollary ideas of self-improvement, body mod-ification, and body disciplining. In the end, globalization does not promote a particular type of body or body discipline, but rather the very idea of body disciplining as a necessary condi-tion of modernity and a necessary aspect of our encounters with the Other. We want to emphasize—and the title of this volume underscores—that we are not arguing that globaliza-tion leads to a homogenization of bodies. Even when similar practices become popular across the globe (e.g., nose jobs or breast implants), the reception of these practices and our understanding of them cannot ignore the local context. Thus, a nose job in Iran can have a vastly different (rebellious) motivation and reception, than a nose job in California (see Karim's essay, chapter 3, in this book).

Similarly, we can look at the practice of female genital cutting (FGC) and its close cousin, female genital cosmetic surgery (FGCS). The two practices have a lot in common. In fact, if one were to compare the description and outcome of a labiaplasty with some of the descriptions of FGC (or female genital mutilation, FGM, as defined by the World Health Organization), the procedures would be unnervingly similar. Yet, much of the Western world bans FGC as a "barbaric" practice. The majority of Australian states, for instance, do not allow even consenting adults to undergo any type of genital "mutilation" or alteration. However, FGCS is not covered under these criminal laws, nor is genital piercing, both of which would fall under the description of genital alteration used in the criminal code (Sullivan 2007). In the meantime, the demand for FGCS is growing and there are increasing reports of girls younger than 14 years, and in some cases as young as 9, undergoing and demanding some form of cosmetic genital surgery (Liao, Taghinejadi, and Creighton 2012). While FGC/FGM is considered one of the worst human rights violations, FGCS is touted as scientific progress, improving women's self-esteem and confidence, and is ultimately seen as women exercising a choice to make their lives better (Braun 2010; Liao, Taghinejadi, and Creighton 2012).[2]

How do we understand this blatant discrepancy in the framing and reception of two practices that are so similar? And what does this illustration tell us about the themes of this book? To put it simply, this example drives home a basic sociological insight: Reality isn't inherent in an object or a practice. Reality is socially constructed and thus must be understood in its proper social context. The example of these practices demands that we look at both global power relations (the Western world vis-à-vis the countries where FGC is practiced), and local power relations (immigrants of color in mostly white Western countries vs. mostly white, privileged women in Western countries) in order to understand why we believe the two practices are so different. Each of the chapters in this book reveals that the context must include the global and the local for us to develop a fuller understanding of beauty practices and ideals in the world today.

*   *   *

GLOBAL *GRINGAS*: WHITENESS IN TRANSNATIONAL PERSPECTIVE

On the crowded street, the young woman becomes separated from her shopping companion. She scans the area, attempting to locate her without success. A street vendor standing nearby has a good view of the block, and tries to get the young woman's attention. "¡Miss Universo!" he cries, not knowing her name.

*   *   *

Beginning at the immigration line in the airport, travelers stop to admire the plump, black-haired toddler with the round gray-blue eyes, and call her pretty. "¡Qué linda!" The fascination continues throughout the subsequent days and weeks. At different times, several people, assuming her to be a fellow citizen, proclaim that she will grow up to be Miss Ecuador. Innocently narcissistic as only young children can be, she revels in the attention, obliging with smiles and winks.

*   *   *

In the United States, where I was born and raised, my appearance is somewhat unremarkable (a state, I acknowledge, made possible by my whiteness). I am average height...well, close enough. I am thin, but not scrawny or malnourished. I don't dress extravagantly, my high school fashion experiments notwithstanding. So I was surprised and amused to be called Miss Universe by a stranger in busy downtown Guayaquil, Ecuador's largest city.

In the second scene above, it is ten years later and I am taking my young daughter on one of many trips to Ecuador, my research site and her father's childhood home. Her face drew comment wherever we went. Her hair was a typical color, length, and texture for this locale. But her pale skin and light eyes made her different and praiseworthy.

*   *   *

Whiteness and beauty are knotted together around the globe. White people enjoy privileges and are often seen as more attractive in my country, too. But the prestige and the attributions of beauty become more striking in a majority nonwhite environment. And white beauty is not just in the face, as with the tot who would be Miss Ecuador, but is associated with particular body types. My thinness and height—I was a head taller than the street vendor and nearly all of our companions on the sidewalk—may also mark me as white. In many places on our planet, the supreme compliment that can be paid to pink-skinned newborns and pigment-challenged potential wives is the recognition of their whiteness. If white is defined as beautiful, then brown skin is ugly and undesirable; and the browner the skin, the more strongly associated with these negative adjectives. These ideas maintain racial hierarchies of bodies in postcolonial societies the world over. Yet the ideas are not monolithic and not always acquiesced to, as the chapters in this volume demonstrate. Individuals shape, and are shaped by, long-standing yet malleable patterns of behavior and public perception. The people we interacted with in these anecdotes are not evil racists; they are experts on what is valued in their society. They do not only judge others, but also themselves. Yet positive evaluations of my daughter's and my (white) appearance are always linked to disapproval of the bodies of others.

Having light skin and stereotypically "Caucasian" features may have led to us being categorized as beautiful in this Latin American setting, but we are not the only women to be judged primarily by our physical attributes. There is a gendered order to everyday life; women are to be looked at, and both men and women do the looking. At the local level, transnational ideas (and ideals) are applied, rejected, or something in between. In the first scene, I may have identified most as a student in that moment, but I was—albeit jokingly—labeled a beauty queen.

My daughter may grow up to be a CEO or a pilot, but she was subtly encouraged to aspire to a more superficial goal. Reducing women to bodies, objectifying them, is a tendency that is discussed in this book's empirical chapters. Rather than simplifying or adopting a strict "anti-" stance toward any of the practices presented here, we hope to reveal the complexities of encounters between *Global Beauty* and *Local Bodies*.

—Erynn Masi de Casanova

* * *

ZOOMING IN AND ZOOMING OUT

Our approach to the study of beauty in globalizing societies is unique because it brings together discussions of practices and ideals at the local level, ideas about the nation, and global or transnational forces and flows. This is an unusual way of studying beauty and bodies, as analyses typically focus on the national level, with fewer examining the local level, and only a handful exploring beauty on a transnational or global scale. Most scholars of beauty—hailing from various disciplinary backgrounds—focus on activities and ideologies (dominant systems of ideas) at the national level.[3] Some of these national case studies have a transnational or global component, as oftentimes national ideas about beauty take place in dialogue with other groups' ideas or preferences. Fewer researchers have taken interaction across national borders or contested ideas of national identity in transnational communities as their starting point for investigations of beauty (Candelario 2007; King-O'Riain 2006). Other researchers keep their ear to the ground, emphasizing variation in and maintenance of beauty practices and beliefs at the local level (Casanova 2004; Cohen et al. 1995). Although it may seem impossibly daunting to present a macrolevel, global view of beauty or social judgment of appearance, a few scholars have undertaken this task (Jones 2010; Nakano Glenn 2009). If we kept the nation as our unit of analysis, this book would simply be an international catalogue of ideas and practices in different countries; as you will see, this is not what *Global Beauty, Local Bodies* is about.

This book, along with its sister volume, *Bodies without Borders*, brings all of these scales—the national, transnational, international, local, and global—into the conversation about beauty, bodies, and embodiment (defined as the subjective experience of being/having a particular body). We would argue that to understand one scale you have to have at least some familiarity with how it connects to the other scales. Local beauty practices are not intelligible without understanding how national identity is conceived of, for example. Although the nation has tended to be the unit of analysis in studies of beauty, the studies presented in this book muddy the tidy contours that outline a country on a map. Beauty cultures (related practices and sets of ideas about bodily attractiveness) can stretch across national borders, and those borders can also contain many different beauty cultures. Modern technology is changing the beauty landscape, allowing individuals, groups, and corporations to circulate images of beauty—or ugliness!—quickly and easily. Mass media companies and corporate advertisers are major players in this diffusion process, and these entities are increasingly transnational as economic globalization progresses in the 2000s.

Although the phrase *Global Beauty* invokes anonymity and homogeneity, individuals (and individual bodies) are engines of globalization. The studies included here do not uniformly find what scholars would call cultural imperialism—that is, beauty ideals and practices moving in only one direction: Global→Local. We can't even draw a bold arrow in the direction National→Local, since the idea of what or who composes the nation is difficult to pin down. We want to disabuse readers of the notion that there is one all-encompassing "Global Beauty" that we can point to as Exhibit A, which is then express-mailed (Or e-mailed! Or Tweeted!) to women and men around the world. Based on the empirical evidence presented here, it makes more sense to think of particular ideas, images, and practices as *mobile*, but without a predetermined direction of influence. Does this mean that power does not exist? No, there are always hierarchies, and some bodies and performances of beauty are circulated more or valued more highly in particular places. The material conditions of people's everyday lives, as we'll see, also dictate to some extent their ability to participate in or reject certain beauty-related behaviors.

We also want to trouble the assumptions associated with the second part of our title: *Local Bodies*. More research is needed that zooms in on what beauty means in people's everyday lives in a particular place and time, while at the same time not losing sight of those other scales (transnational, national, etc.). With this two-word phrase, we do not mean to imply that all bodies that share a geographic location conform to the same norms of beauty and attractiveness. In fact, some of these chapters present evidence that even in very small local areas, radically different conceptions and practices of beauty may coexist. Our idea of which bodies count as local bodies is an inclusive one; we see bodies not as immutably fixed in geographic locations, but as moveable and adaptable. Internal and international migration bring previously distant beauty practices into proximity, for example; this is another reason that a strictly national frame is inadequate in an era of unprecedented globalization. Whether or not people are recognized by their neighbors as "locals," they are ultimately performing styles of embodiment, and working to make their bodies beautiful, in specific local settings. No matter the scale at which their research inquiry begins, our authors are adept at moving back and forth between micro-, meso-, and macrolevels in their explorations of beauty and globalization.

SCOPE AND PLAN OF THE BOOK

What started out as a book project eventually turned into two books: *Global Beauty, Local Bodies* and its sister volume *Bodies without Borders*. Though there is overlap between the two volumes, we had enough chapters focusing on beauty alone to merit their own book. This is not surprising. The discussion of the body has often focused on beauty ideals. It is also not surprising that so many of our chapters focus on women. Although scholars are slowly turning their attention to men's bodies and embodiment, it is interesting to note that a discussion of men's bodies often happens in stereotypically "masculine" domains. In this volume we see a focus on male athletes, for instance. Although we were able to capture a remarkable degree of geographical diversity in this collection, our inability to include a variety of chapters on men's bodies and embodiment reveals the skewed nature of the study of the body and embodiment. These are noticeable gaps that need to be filled as we move forward in our study of the body.

The chapters in this book cover an extraordinary range of diverse views and topics while focusing on beauty. The first chapter, by Oluwakemi M. Balogun and Kimberly Kay Hoang, takes an innovative comparative-ethnographic approach, emphasizing how women play a significant role in constructing new forms of embodied femininities that help to position their nations, Nigeria and Vietnam, as the "giant" of Africa and the "dragon" of Asia, respectively. Drawing on extensive fieldwork on Nigerian beauty pageants and Vietnam's sex industry, the chapter illustrates how women in these industries create new constructions of beauty that actively contest the dominant Western ideal.

Susanne Hofmann discusses aesthetic labor in Tijuana's sex industry. The women sex workers—most of whom are internal migrants from rural areas of Mexico—invest time, money, and effort in constructing an image or persona to fit clients' fantasies and desires. In this highly competitive sex work location, the readers learn how sex workers deploy a variety of strategies in order to be successful.

Next, Persis M. Karim offers the readers a peek into the politics of cosmetic surgery, in this case, a nose job, among Iranian immigrants in the United States. In an insightful and humorous essay and poem, Karim pushes readers to acknowledge how one's location changes the meaning of an act and discourages simplistic conclusions regarding bodily practices.

Michelle Newton-Francis and Salvador Vidal-Ortiz take us into the world of Hooters in Bogotá, Colombia. In the United States, the Hooters Girl is constructed in a manner consistent with a particular kind of femininity and sexuality that is appealing to the national audience—the "cheerleader" or "surfer girl next door," as our authors argue. But, what happens when the Hooters restaurant chain becomes globalized? Can it sell its version of femininity and sex appeal to the customers in Colombia? Once again, our authors reveal the fissures in the process of globalization and emphasize the importance of local context by discussing the adaptations that the Hooters Girl must go through in order for the restaurant to be a success in Latin America.

The next chapter is by Kaija Bergen, who offers a personal reflection piece as a white woman located in Cambodia, who is admired for her "beautiful" white skin. Bergen's discussion also points to the importance of location in understanding beauty ideals and beauty practices. It is only through a "dislocation" that Bergen becomes aware of the arbitrariness of beauty ideals (shaving leg hair, for instance) and begins to question those ideals and practices that are seen as "normal" or universal in different settings.

Building on the theme of dislocation, Lucy Lowe's chapter takes us into the vivid world of Somali immigrants in Kenya. Drawing on her extensive fieldwork in the Somali-dominated neighborhood of Eastleigh, Nairobi, she explores how beauty was perceived and achieved among women as a site of agency through which they could attempt to secure their futures. Furthermore, the chapter examines how consumption of "foreign" beauty, from Nairobi fashion to Indian soap operas and American music videos, influenced migrating concepts of beauty, gender, and the reproduction and continuation of the Somali nation through the bodies of diaspora women.

Jaita Talukdar's personal reflection on body weight, first in India as a teenager and then later on living in the United States, poignantly reveals one of the central themes of this volume—our ideas of modernity and progress are often tied to our notions of the ideal body. As the readers experience the disciplining of the body through this personal essay, they realize that this is not simply about vanity. A modern body, especially one that wishes to be a part of the new global world, must look a certain way.

The last chapter in this volume, by Jan Wickman and Fredrik Langeland, discusses the variation in the definitions of metrosexuality by analyzing contrasting examples of sexualized media representations of athletes from three Nordic countries: Finland, Norway, and Sweden. Again, we see the intersection of the global and local, as the authors discuss the global phenomenon of metrosexuality, but underscore the importance of the national context and national identity while analyzing images of male athletes.

This book and its companion volume, *Bodies without Borders*, have been in the making in our minds for a long time. In our teaching and research, we had been surprised to find that scholars of globalization have neglected to study the body more deeply and that scholars of the body and embodiment have neglected to study the implications of globalization on this most intimate aspect of our lives. But when you study the flow of people from one place to another, the flow of ideas from one place to another, and global power relations closely, you cannot escape the knowledge that our bodies are deeply embroiled in the processes of globalization. Yet, as these chapters show, globalization is experienced and embodied in particular ways, in specific locations. Thus, just as our understanding of the local is incomplete without an examination of the global, we must never lose sight of the local in our examination of the global. The chapters in *Global Beauty, Local Bodies* initiate a long overdue discussion that engages the local and the global simultaneously.

NOTES

1. Colorism, that is, skin color-based prejudice and discrimination, is not the same as racism. Colorism can exist among people who belong to the same "race," but is not found only among non-white populations. In majority-white societies such as the United States, whites also typically differentiate minorities based on how dark- or light-skinned they are, which has implications for individuals' life chances (Hunter 2002). Whether practiced by people who belong to the same racial group or not, colorism is focused on the shade of one's skin rather than racial classifications.

2. For a longer discussion on this topic see Afshan Jafar's TedxTalk (2012): http://www.youtube.com/watch?v=BaxnvwffWbE.

3. For examples, see Banner (1983); Dewey (2008); Edmonds (2010); Peiss (1998); and Rahier (1998).

1

REFASHIONING GLOBAL BODIES

COSMOPOLITAN FEMININITIES IN NIGERIAN BEAUTY PAGEANTS AND THE VIETNAMESE SEX INDUSTRY

Oluwakemi M. Balogun and
Kimberly Kay Hoang

WOMEN'S BODIES AS NATIONAL REPRESENTATIONS

Women's bodies are symbolic sites where debates about the development of a nation take place. Shifts in the global economy, cultural globalization, and postcolonial trajectories map onto women's altered embodiments (Dewey 2008; Mani 1998; Otis 2012). These bodies represent a nation's shift toward modernity (Rofel 1999) through economic progress and development. Bodily practices and markers of appearance such as dress, makeup, and grooming are vehicles of collective identity in which women's bodies are often the terrain where national identities are produced, maintained, and resisted (Choo 2006; Gal and Kligman 2000; Huisman and Hondagneu-Sotelo 2005). The embodiment literature has established how cultural constructions of the body are shaped by the role of local and global media consumption (Casanova 2004), the quest for upward mobility (Edmonds 2010; Rahier 1998), and the tension ethnic and racial minorities face in establishing ethnic or racial authenticity while incorporating into multicultural societies (Craig 2002; King-O'Riain 2008; Rogers 1998).

Several studies claim that Western-defined beauty standards such as lighter skin and slim figures have spread throughout the world as a result of the multinational cosmetics industry (Chapkis 1986), diffusion of mass media (Shilling 2003), and the production of New York and Paris as the fashion capitals to a new generation of consumers around the world (Jones 2011). Other scholars dispute the Western origins of these beauty standards, noting that in some instances they may predate colonialism and have a much more complex internal history unconnected to the West (Li et al. 2008; Wagatsuma 1967). This debate highlights the extent to which beauty standards are internally or externally constructed within a nation in an era of rapid globalization. The current scholarship often assumes a one-dimensional understanding of the diffusion of Westernization and ignores how developing countries fuse embodied practices and nation-building projects to emerge into the contemporary global economy.

We bridge both of these perspectives by centering the intricate local, national, and global forces at work in the body projects of Nigerian beauty contestants and Vietnamese sex workers. In this chapter we ask: How do women's bodies come to shift nation-based hierarchies to represent their nation's rising status in the global economy? Nigeria and Vietnam both filter their beauty ideals through the prism of international standards. These standards are perceived to be a set of principles that all nations around the world are held accountable to, but are shaped by local context. International standards serve as buzzwords that frame not only beauty culture, but also the political economy. In Dewey's study (2008) of Miss India, she notes that economic liberalization policies shape how the pageant seeks to conform to international standards while preserving Indian national culture for a global audience. Our participants view these standards as being previously defined by the West, linking this control to an imperialist past and economic might. We show how the women in both industries recognize, and take part in, the shifting orientation of international beauty standards. Women in both nations perceive this shift in international standards of beauty as trending toward the ascendance of developing countries that are beginning to have a presence on a global stage. That is, they counteract assumptions of Western dominance within global beauty culture by emphasizing that international beauty standards are multidirectional. Developing countries are internally redefining their own beauty standards, which, while not completely dismissive of the role of Western influence, center on contemporary African and Asian ideals. Simultaneously, they envision that in the near future their countries will play a significant role in defining these international beauty standards for other nations, which they directly link to their establishment as leaders in a globalizing world economy.

GLOBAL CONNECTIONS: TWO DIFFERENT COUNTRIES AND INDUSTRIES

We now turn to our ethnographic data, independently collected by each author in Nigeria and Vietnam. In Nigeria we examine the country's premier national beauty pageant, where Oluwakemi Balogun conducted 10 months of ethnographic field research and 55 formal interviews with female pageant contestants, organizers (who are mostly men), corporate sponsors, and judges. Turning to Vietnam, we focus on the highest paying niche market of Ho Chi Minh City's (henceforth referred to as HCMC) sex industry, where Kimberly Hoang completed 15 months of ethnographic field research along with 54 informal interviews with female sex workers, male clients, bar owners, and madams. Both studies took place between 2009 and 2010, allowing us to observe comparable contemporary patterns in two countries where women played a critical role in representing their respective nations' *shifting* place in the new global economy.[1] These shifts discursively map onto the representations of women's bodies. These bodies highlight the adoption and redefinition of international standards of beauty that do not always neatly align with Western hegemonic ideals.

Vietnam and Nigeria are both emerging nations. Following in the footsteps of capitalist economies in their respective continents, they are both the second fastest growing economies in their regions. Hayton (2010) dubs Vietnam the rising "dragon" following in China's trajectory of rapid economic development. Similarly, Nigeria is lauded as the up-and-coming "giant" right after South Africa. These two seemingly disparate nations have

one thing in common: they are both rapidly developing economies that provide people with new opportunities to reconfigure
social structures and the place of their nations in the global
world.

Much of the scholarship on economic restructuring utilizes
a top-down analysis focused on the state or on the movement
of capital in analyzing how developing countries modify their
place in the international political economy. However, this
study takes an ethnographic approach that emphasizes how
women in local spaces play a significant role in constructing
new forms of embodied femininities in the cultural economy
(Mears 2010) in order to reposition themselves—and by extension their nations—on the global stage. Both beauty pageants
and the sex industry are sites of heightened femininity that serve
as fruitful spaces to study how women interpret and understand
the political and economic changes that are taking place in their
countries. Our complementary data focus on the distinct *looks*
cultivated by women in both institutions, which remain managed in part by others around them, such as madams, patrons,
fans, and organizers. We reveal the public *and* hidden constructions of femininity through bodily practices within cultural performance and the underground economy.

Beauty Pageants in Nigeria

The Silverbird Group, a Nigerian-based media conglomer-
ate with branches in Ghana and Kenya, coordinates the Most
Beautiful Girl in Nigeria. The Most Beautiful Girl in Nigeria
(MBGN) is the most visible beauty pageant on the Nigerian
national scene and has sent contestants to the top beauty pageants in the world since 1986. At the finale, MBGN chooses
five winners who go on to represent Nigeria at different beauty,
modeling, and promotional contests within the country and
around the world. The winner and first-runner up continue
on to Miss World (a British-based pageant) and Miss Universe
(a US-based pageant), respectively. MBGN produced the first
black African winner of the Miss World contest in 2001.

The organizers of this pageant push forward a gendered
nationalist vision that molds women into cosmopolitan subjects

who publicly represent the nation. While its organizers are primarily Nigeria based, MBGN also relied on grooming and production experts outside of Nigeria, most notably from the United States and South Africa. These experts are said to internationalize the Nigerian pageant system and offer contestants a leg up in the Miss World and Miss Universe competitions. Contestants are trained to embody "international standards" through their looks, movements, and high self-confidence. The Silverbird Group has sponsored three different international pageants within Nigeria, which brought in delegates from throughout the world to compete: the Miss Intercontinental pageant from 1986 to 1990, Miss World 2002, and Miss Silverbird International in 2004. Contestants are chosen at various cities throughout Nigeria, primarily in the southern part of the country (Benin and Port Harcourt in the Central South, Lagos in the Southwest, and the capital Abuja in 2010). Semifinalists selected from these screening venues then compete in Lagos, where 30 finalists are chosen to compete in the final show. The pageant's finale activities are usually centered around Lagos, the port-metropolis and commercial hub, where much of the country's wealth is concentrated.

Pageant contestants' public exposure and fame provided them with direct access to some of the nation's most powerful political officials, elite businessmen, and acclaimed celebrities. MBGN's first prize package includes: $21,000, a brand-new car, endorsement deals, and lavish gifts. Through shifts in embodiment, such as access to high-end styling treatments and brand-name clothing, contestants (especially pageant winners) embrace a newfound lifestyle that signals the nation's rising position in the international political economy.

High-End Commercial Sex in Vietnam

Khong Sao Bar, located in the heart of HCMC's business district, services the highest-paying niche market for commercial sex in Vietnam. As one of the most profitable bars in HCMC that caters to the country's wealthiest businessmen and political officials, this bar was hidden. In order to get a table, the clients had to have a preexisting relationship with the madam or be introduced by a top-paying regular client. High-end Vietnamese sex workers entertain foreign guests, helping to cement business deals that direct capital into the country while also serving as emblems of progress and development.

Vietnam's elite businessmen operate some of the nation's top finance, real estate, and trade companies. Collectively, these men are responsible for directing the majority of capital through foreign direct investment (FDI) projects into the country. As regular patrons, they spend an average of $1,000–2,000 per night and $15,000–20,000 per month in *Khong Sao Bar*. There were three madams who ran the bar. These women trained sex workers on how to sit, drink, sing, dance, and maintain appropriate relationships with their clients. Workers earned roughly $2,000 per month, comprising tips for joining men at their tables and $150–200 for each sexual encounter. The madams earned $3,000–4,000 per month in tips and got a small percentage of all alcohol sales in the bar.[2] In order to sell Vietnam to foreigners as a place to do business, local firms rely on hostess bars to help them dramatize Vietnam's potential as a lucrative location for foreign investment. In sharing the ritual of male drinking (Allison 1994) with their guests, clients showcased their wealth. They also noted the transformation of the nation through a focus on the changes that women's bodies underwent. Sex workers embodied transformations through skin lightening creams, plastic surgery, and conspicuous consumption that helped local Vietnamese men convey to their investors a lived sense of Vietnam's dynamism and faith in the nation's promise for rapid economic development.

SITES OF ATTRACTIVENESS: REDEFINING "INTERNATIONAL STANDARDS"

In both pageant contests in Nigeria and the sex industry in Vietnam, women had to contest with converging local, national, and global idealized femininities. Many of these women engaged in embodied practices that on the surface appear to embrace Western standards, yet they defied this interpretation by framing their own practices as way to solidify their cosmopolitan status. Similarly, Saraswati (2010) argues that beauty practices like skin lightening, which are usually associated with whiteness, are not automatically equated with Caucasian ideals, but rather with cosmopolitanism and transnational mobility. Women in both countries attempted to pluralize and expand international standards of beauty, sometimes with critical reflections on Western feminine ideals.

The MBGN pageant strived to carefully craft a glossy image of their contestants to represent the very best of the nation as a whole. Pageant organizers insisted that the contestants strived toward fulfilling international standards of beauty, which they did not equate with universal Western-based criteria. MBGN defined international beauty standards through a set of criteria focused on height, weight, body shape, and facial features. While they acknowledged that these criteria were previously dictated by American and British definitions due to their control of international pageants, they also insisted that these pageants have been forced to change as a result of the increasing participation and success of contestants from the developing world. They emphasized the multidimensionality of international standards by strategizing within them, selecting delegates who tapped into distinct niches in global pageantry. For example, candidates sent to Miss Universe are described as glamorous "model-types" who are tall, slim, and dark. Miss Universe, owned by the mogul Donald Trump, is viewed as a corporate enterprise focused on integrating the modeling industry into the Trump business empire. Dark skin was viewed as important for Africans because it makes them stand out as exotic. In contrast, organizers sent a "girl next door" type to the Miss World pageant, a British pageant organized and privately owned by the Morley family. To appeal to Miss World's "beauty with a purpose" tagline, organizers and judges focused on picking a fresh-faced, innocent-looking candidate with mass appeal. These candidates tended to be more shapely, shorter, and lighter-skinned.

These parallel preferences for light-skinned and darker-skinned contestants were equally elevated as part of the ideal "international standards" for Nigerian representatives. Beyond highlighting the social dimensions of skin color, by both marketing light skin as a marker of global "mass appeal" (Glenn 2009) and capitalizing on dark skin as a form of desired "exotic beauty," MBGN manages beauty ideals through a global cultural economy, highlighting some flexibility in striving for international legitimacy. The national director of MBGN detailed the difference in the following way:

> Well for Miss World, it is a family-owned organization. They are looking for a likeable personality in a woman, someone they would consider a daughter. A nice person, sweet and lovable person who can achieve goals. For Miss Universe, I see them looking for an exotic model who can model for Gucci and the rest, you know, a high-flying person. We will be looking for a wholesome person for Miss World and a commercially viable person for Miss Universe. Someone that can do an advert, be on the billboard, can stand in front of the TV camera.

MBGN Organizers contrasted the two international pageants as striving toward a natural versus a glamorous look, which they mapped onto skin color. That is, through this strategy of emphasizing the marketability of skin color, pageant organizers viewed international standards in a much more flexible manner that did not *always* value light skin.

Contestants worked to achieve vibrant, even-toned skin free of blemishes, regardless of their skin color. Visible scars were pointed out, scrutinized, and debated over the course of the screening process. During the audition, a couple of the chaperones pointed out a woman who had tried out for the past two cycles of MBGN. While she had made the cut in the past, she did not go on to win the crown. "Each year she comes back cleaner and cleaner," one chaperone said. When asked to clarify what she meant by this statement, she responded that each time she returns to audition her skin looks fairer. This observation was just one of many which pointed to how contestants' bodies physically shifted as a result of participating in the pageant, which in this case was linked to access to exclusive skin and makeup treatments that seemingly made their skin "cleaner." In preparation for the pageant, contestants often rubbed skin ointments and lightening creams to "tone" or smooth out their complexions. In particular, contestants sought to eliminate black patches on their elbows and knees. These beauty ideals were not framed in terms of achieving Western ideals, but rather as a means of establishing class status, since lighter skin was not always the desired result, but rather achieving an even skin tone. This finding is similar to other scholars' work, which shows how lighter skin serves as a means of not only promoting Western beauty ideals, but also secures and verifies upward mobility (Hunter 1998; Pierre 2008; Rahier 1999). The ability to physically alter one's body and the associated financial resources attached to these shifts denoted a higher-class status.

The pageant often touted its winner's success in becoming Miss World 2001 as a signal that international standards were shifting toward the recognition of Africa. Shortly after clinching the title, Agbani Darego, Nigeria's representative to the Miss World pageant, gushed, "I have made history...black is beautiful." While many Nigerians were critical of Darego's win, complaining that she did not conform to Nigerian beauty ideals of shorter heavier-set bodies, others also noted how the trend for *lepa* (slim) tall bodies highlighted new definitions of youth, wealth, and health in the country. The latter perspective stressed how preference for more voluptuous bodies signaled Nigeria's traditional past. In the more than ten years since Darego's win, MBGN moved from a screening process that emphasized very precise measurements (everything from neck circumference to hip to ankle lengths) to a much more open system, which they viewed as still competitive. Pageant participants viewed Africa as the next frontier for beauty, as one MBGN organizer stated:

> If you are looking at [international] pageants today, Americans never win...I will tell you why: because Americans are so programmed from birth, they are so mechanical that there is no natural beauty in an American woman today. Now the world is looking for natural beauty and guess what the future is? Africa...African beauty is becoming exciting. All of a sudden we are becoming the future, and we could win every year, but we've got to stay focused.

They claimed that beauty queens from Western countries were plastic-looking, unusually pin-thin, and plain. In contrast, African beauty was seen to be more natural and striking. Through these tactics, MBGN signaled the new direction of international pageantry, which increasingly focused on members of the developing world like Venezuela and India as necessary to the growth of the market (Dewey 2008; Ochoa 2005). Exotic looks, dark skin for example, provided Nigerian beauty contestants with an edge internationally.

A similar focus on skin color is evident among Vietnamese sex workers. In *Khong Sao Bar* women worked hard to lighten their skin. However, they claimed that they had no desire to look Western. In fact they worked deliberately to de-Westernize their bodies. Nhung, for example, said, "in the past, everyone wanted to look Western, but that is old (*sen*). Now...the new modern (*hien dai*) is Asian." In this bar, looking "Western" was not synonymous with looking modern or following an international standard of beauty. In fact, Western women were considered unattractive because they were too overweight, wore clothes that looked messy (*bay hay*), and looked too masculine. For instance, Huong explained, "[They] look like men, [with] squared bodies and saggy boobs. Asian women have smaller bones, smaller waistlines, small hips, and boobs that fit their bodies. When you are smaller, you look gentler, softer, and more feminine."

Instead, the workers wanted to look like the women from Hong Kong, Korea, or Japan. Local sex workers began to modify their bodies to conform to international, Asian standards of beauty as Asia. Imaginations of the *modern girl* (Weinbaum et al. 2008) from fully developed capitalist societies within Asia began to gain a stronger presence in the international global economy. Women lightened their skin, accentuated their eyebrows, and worked to look like Korean pop stars. When business was slow, the women sent off the male bar attendants to purchase Korean and Japanese magazines from street vendors so they could learn the latest styles. In contrast to the Western magazines like *Vanity Fair*, *Vogue*, and *Cosmopolitan*, which were available in the local market but nearly absent in the bars, Korean and Japanese magazines provided women with idealized feminine bodies that they sought to emulate. When asked why Korean and Japanese magazines were so popular, the women explained that the models in Korean and Japanese magazines had smaller frames and bodies that were more petite compared to the models in Western magazines. Asian magazines also had lengthy articles about the various kinds of skin care products and cosmetic surgeries developed and produced specifically for Asian women. As Thuy, a 24-year-old sex worker, explained, these surgeries and skin care products are designed to bring out their "natural" beauty and enhance their Asian features. As in other parts of Asia, unsuccessful surgeries are often defined as producing an unnatural or Western appearance (Holliday and Elfving-Hwang 2012).

In conversations with workers about trying to look Korean or Japanese, many of them talked about how they believed the global center was shifting away from the West and toward Asia. Blonde hair and blue eyes were desires of the past. Sex workers pointed to the Dream Girls, a famous group of five singers from Korea, as their ideal of femininity. The irony behind this way of contrasting Asian versus Western bodies is that the Dream Girls all had reconstructive surgery to build nose bridges, construct double eyelids, and lighten their skin, changes associated with Western standards of beauty. On the surface, the Dream Girls' reconstructive surgeries seem to emulate Western standards of beauty, however, Dai a 19-year-old hostess worker critically explained the difference:

> When women use skin lightening creams for the face and body people think they want to look like white people in America or Europe, but actually the true skin color of women in Asia is white. When a baby is born in Japan, Korea, or Vietnam what color is their skin? It is fair and white, right? Dark skin is from going out in the sun a lot. We are just trying to bring out our natural beauty...No one wants to look Western here anymore. In the West the women, even the models, are a lot fatter and their bodies are much more squared. They look hard instead of soft and feminine the way true Asians do. People come to Asia for beautiful Asian women, not for women who look Western.

Dai's clarification points to an international standard of beauty that is much more nuanced than one associated with Western ideals. Beyond skin color, Dai's comment also illustrates the attempt to resist Western ideals and highlight ideals that are regionally specific, and in this case distinctly Asian. While one might interpret these bodily modifications as Westernizing practices in the East, *localized discourses* that intersect with more globalized practices of cosmetic surgery (Holliday and Elfving-Hwang 2012) help to explain why such practices cannot simply be understood in terms of Westernization. This deliberate focus on aspects of Western standards of beauty that the women work to avoid illustrates how local, regional, and global ideals converge in women's practices.

The focus on skin tones in both Nigeria and Vietnam illustrates a shift in international standards of beauty as they intersect with local and regional ideals. While these practices might easily be read as Western, our data complicate this framework. Participants do not view beauty ideals in such a one-dimensional way. Rather, they see new beauty standards as striving toward an international ideal in which regions like Asia and Africa are beginning to have a greater influence. The point here is not to dismiss the cultural influence of Western idealized femininities, but rather to provide some nuance to emphasize how international standards are evolving toward centering African and Asian beauty ideals as prominent. In Nigeria this involves redefining dark—albeit unblemished—skin as exotic, unique, and beautiful, while in Vietnam the cultural ideals are shifting away from the West and toward an ideal type based on East Asian pop stars. In this way globalization creates new possibilities for alternative standards of beauty, whether Asian or African, that both emulate and rival Western standards of attractiveness.

WOMAN AS NATION: SYMBOLIZING ECONOMIC PROGRESS IN NIGERIA AND VIETNAM

In both Nigeria and Vietnam, women's economic mobility and the bodily transformations that come with it represent the nation's economic prosperity or progress, both to the country and around the world. Transforming women from "country bumpkins" into "cosmopolitan subjects" was crucial to understanding the nations' shifting position in the global economy. Shifts in embodiment that are associated with a modern look were routinely noticed and celebrated by the women themselves, as well as those around them, as a means of highlighting economic mobility.

MBGN signals the rapid upward mobility of its contestants by focusing on the process of "grooming" that they undergo during camp (the pre-show training and rehearsal period). The training period was noted as having a profound impact on both the demeanor and physical embodiment of contestants, which was directly linked to the class mobility of contestants. People would constantly comment that contestants would change over the course of the camp period that led up to the finale. For example, one of the chaperones, Ada, motioned toward the group of contestants gathered outside the pool of the 5-star hotel that served as host for camp, "They will all change. You'll see them next year and you won't even recognize them." With access to hairstylists and makeup artists provided by MBGN and the opportunity to interact with some of the best Nigerian fashion designers, contestants' physical embodiment and, by extension, their lifestyles, were expected to change over the course of the contest and beyond.

All contestants underwent a hair consultation and makeover during the course of the competition. While at earlier stages of the competition there was some variety in hairstyles from shorter buns to long micro-braids, by the time of the final show there was a great deal of uniformity among contestants. Most wore waist-long weaves or lace-front wigs in elaborate upswept hairstyles or down in soft waves to showcase their silky-long hair texture and length. Contestants spent hundreds of dollars on "premium" hair from India, Peru, or Brazil. It was not uncommon for contestants to compliment each other on their hair and ask, "Where is it from?" as a way of tracing the origins of the hair to measure each other's competitiveness and level of financial resources. They were also expected to own the latest Blackberry phones, designer handbags, and platform heels. Contestants were to eschew secondhand and counterfeit brand-name clothing and accessories, which flood the Nigerian market, and instead wear Nigerian couture and shop at malls with trendy American or European labels.

Joy, the 2009 beauty queen, entered the room wearing low-heeled strappy sandals and a black sequined dress, her crown nestled inside a large silver case. She was being interviewed and photographed for a magazine article. The photographer introduced himself. "We have met before—I don't know if you remember me. Wow! You've really changed, oh!" he remarked, looking her up and down. "Silverbird's money has changed me," Joy responded with a smile. What others pinpointed as the core change that the contestants undergo is the cultivation of a new polished image, which directly translates into increased social status and a newly acquired jet-set lifestyle. On the day before the finale, during the dress rehearsals while all the contestants practiced onstage, Mr. Oke, a staff member at Silverbird, commented, "I'm always scared of these girls. They are powerful. That's why I'm always nice to them. They're all going to dump their boyfriends after this is over. You'd be surprised, one of them might be the future wife to a minister [head of government ministry]; they might just be the one to make that phone call to make or destroy a deal." Mr. Oke's reference to contestants was meant to signal their acquired access to the political elite, well connected to the international business world. Organizers stressed how contestants moved into new accommodations (usually on the Island),[3] gained access to chauffeurs, and consumed the trendiest luxury brands.

In addition to the economic mobility that contestants acquire, the pageant opens doors for women to make connections with political officials and business elites, because these women served as cultural ambassadors representing their nation. Pageant contestants stressed their own symbolic role, in which showcasing their own positive attributes served a larger function of highlighting the good elements of Nigerian society. Faith, a pageant contestant, likened beauty pageants to football and Nollywood (Nigeria's film industry), which she viewed as cultural products that could be exported to other countries to cement relationships:

> Pageantry is another means of promoting the good side and nature of Nigeria. In other words, it means getting the attention of other countries down here to Nigeria, showing them how peaceful, hospitable we are here in Nigeria. You know, create a relationship with other countries. Pageantry is one of the ways that can be achieved.

Faith understood her position and that of other contestants to be that of cultural ambassadors who could use their beauty, charm, and hospitality to gain attention for Nigeria and maintain ties to other countries around the world. Pageant contestants also perceived pageants as a vehicle to rebuild negative perceptions of Nigeria. Camryn, MBGN's queen during the 2010 cycle, noted how pageants could address foreigners' views of women's issues in Nigeria:

> Number one, we are telling people that here in Africa, women are also given a sense of responsibility, and pageantry has been able to say that. There is a whole lot of misconception about women in Africa, in Nigeria; [that] they are treated badly, in terms of widowhood, sex education and all of that. So pageantry has been able to break that barrier...Pageantry has really done well; it has been able to not just help Nigeria but her citizens.

Beyond their work as cultural diplomats who sought to rectify the negative image of Nigeria abroad and build connections to other nations, contestants also believed that beauty queens needed to reach out to the community (especially youth) and use their titles to gain access to political leaders in order to serve as liaisons between the general public and the state. Doyin shared:

> When you have the crown, it's like an open door for you, so it is left to you to keep the door open, it is left to you to carry on the legacy and show them that I'm actually capable, I'm beautiful, I've got the crown, I've got brains. With the crown, you could enter the National Assembly, you could go to visit the governor, he'll give you a listening ear, he will say oh yes, she is a beauty queen and she has something she is doing.

Similarly, Penelope noted that while pageants provide you with a platform, it is up to the individual beauty queen to use her voice effectively to "speak for the people," push the country forward, and provide a collective voice for the nation:

> I always say that pageants gives you a voice and then you speak...for me it is not just about the pageant...it is about how she's helping the country, how can she take the message [forward], how is she using that crown on top her head? It gives you a voice to speak for the people.

Through these statements, both Doyin and Penelope emphasized the ways that women came to represent Nigeria both
within the country and to the rest of the world. Beauty queens
saw themselves as national figures and role models who were
under constant scrutiny in the press and within the court of
public opinion. Contestants' jet-set lifestyle, presented in part
through their shifting bodily practices and consumption ideologies, shaped their ability to present a modern nation to the
world and aid in national development.

In Vietnam's high-end niche market, sex workers emulated
new international standards of beauty as a means of differentiat-
ing themselves from rural women whom they viewed as poor,
backward, and unsophisticated. Several of the madams worked
hard to build a network of young beautiful village women whom
they could teach to reconfigure their bodies to look like cosmopolitan subjects. In order to feel desirable and to *be* desired
by men, workers engaged in a variety of disciplinary practices
to transform themselves from village "bumpkins" into modern
urbanites. The highest earners were the ones who invested in
plastic surgery. Hanh, one of the madams, advised her workers:

> When you are new, it's better to invest in cheaper dresses and
> save your money, because men will bring you into their tables
> because you are a fresh face. After you've been here for a couple
> of months you need to do things to stay fresh (*tuoi*). Like with
> Diem, I told her to take 300 dollars and get a nose job. After
> she got a nose job, men pulled her into all of their tables. They
> wanted to see her new face, her change. She went from looking
> like a poor village girl to looking more modern (*hien dai*). Men
> do not come in here to sit next to village girls—they come here
> to sit next to modern women.

Hanh shared that she much preferred to work with girls from
the village, because women from the city and high-end models
were much more resistant to undergoing risky bodily transformations. She said:

> I try to find the most beautiful women from the villages because
> I can transform their bodies more easily. When I tell them to get
> a surgery, they listen to me and they don't argue. Men come in

here for a certain look; they want to show us off to their guests from around the world.

While the madams play an important role in shaping women's bodies, so too do the clients—their desires ultimately bring business into the bars, and when women know how to appeal to clients' desires they often make more money. For example, while sitting at a table with Diem a few days after she recovered from her rhinoplasty operation, Quang, a 39-year-old client, pointed to her nose and asked everyone at the table, "What do you think of her nose? She doesn't look like a poor country girl anymore, does she? This face looks modern!" Everyone laughed as he kissed her nose, raised his glass and cheered everyone at the table. Diem shyly covered her nose with her fingers because the bruising had not completely healed. As she looked down, the men complimented her and told her that the bruising was evidence of economic progress because it meant that she could afford to have surgery.

For these men—most of whom are connected to the foreign direct investment projects channeled into the country— Diem's nose was a sign of the nation's progress because it demonstrated that even the poorest women in Vietnam, who came from rural families, were able to capitalize on the nation's changing position in the global economy. Plastic surgery was no longer something only the rich could afford; it was something that poor rural women now working in the sex industry could afford. The technologically altered bodies became symbols to elite businessmen that reflected the nation's economic progress. Several clients even paid for these surgeries as a gift to new women with whom they enjoyed sitting. Phuong, a 23-year-old worker in the bar, explained:

> All of us come from the village and it takes time to get the look that makes money. If the clients think that you have potential to make them look good in front of their friends, they will pay for your surgeries as gifts. I got my first job done when I was 19—that guy paid for my nose bridge. A year later another guy paid for me to get eye surgery and someone else paid for my breast implants. These surgeries are expensive, and some girls

work for months just to save enough money for them, but if they were smart they would let the guys pay for it.

Doctors trained in Singapore, Thailand, Korea, and the United States who know how to work specifically with Asian women's faces and bodies are the ones who perform nearly all of these procedures. Many of the women expressed concerns about going to doctors who work mostly on Western women for anything other than weight loss surgeries, because they worried that those doctors would not know how to enhance their natural beauty through subtle bodily modifications on their noses, eyelids, and breasts. Doctors who have a great deal of experience in Asia know how to enhance women's bodies according to regionally bound standards of beauty that conform to shifting international standards while simultaneously preserving their national culture (Dewey 2008).

These bodily modifications, which highlight the women's malleability, mobility, and modernity, were crucial to local Vietnamese business elites because they signaled the nation's progress and economic development. They were so important that it was not uncommon for a male client to purchase these procedures as a gift to the sex workers in the bar. When some of the clients were asked why they paid for such services, Dong, a 60-year-old businessman, explained:

> When you look from the outside in, it seems like they need our money, but we need them just as much as they need us. When you bring in businessmen from Asia, you can say "look, this country is growing and developing so much that even the poorest village girls can afford to get plastic surgery." It shows them that we're a nation that is growing very rapidly and there is a lot of potential in our market. They represent Vietnam to the most important people, our investors!

Sex workers' altered embodiments help to transform the ways that investors see Vietnam. These bodies provide foreign investors with a sense of hope and faith that their investments will have lucrative returns because Vietnam is a nation on the move, where even the poorest of the poor are beginning to reap the rewards of economic development. Following the 2008 global

economic crisis, the West was no longer the primary source of
foreign direct investment in Vietnam. Instead, Asian countries
like Korea, Japan, Malaysia, Singapore, and Taiwan were the
main investors (GSO Vietnam 2011). This shift is carefully
mapped onto sex workers' bodies as they engage in practices
that make them look like women in fully developed capitalist
societies within Asia. Over time, the women began to develop
and maintain a certain *look* in the bar, one that hinged on an
international standard of beauty that was not necessarily synonymous with a Western ideal, but rather converged with regional
ideals within Asia.

Both the sex work and beauty pageant industries are shaped
by the same broader economic forces, such as rapid local development, the global growth of "frontier markets" seeking to be
the next major economic players, and the emergence of a homegrown super-elite who remain plugged into the international
political economy. They also both cater to a foreign audience
who they do not view as uniformly Western. Both cases note a
visible shift in the displacement of once-powerful countries. In
the case of Nigerian beauty pageants, beauty contestants view
developing countries like Venezuela and India as their main
rivals, and view countries like the United States and England
as less competitive. Contestants' competitiveness in international pageants helps showcase the positive attributes of the
country to the world. Moreover, their association with members of the Nigerian political and business elite highlights the
nation's upward trajectory and wealth. In Vietnam, the changing source of foreign direct investment capital from the West to
Asian markets translates into a new group of homegrown elites
and Asian businessmen who serve as clientele in high-end bars
like *Khong Sao Bar*. For Vietnamese sex workers, satisfying the
needs of their Asian clientele helps reinforce the growing status of the region. In striving toward embodiment ideals, which
they view as *international* and regionally specific rather than
Western, they note the recognition of previously marginalized
countries in the global arena. While these beauty practices do
not entirely dismiss Western influence, participants themselves
interpret them through an alternative lens that recognizes difference within international standards.

REFASHIONING COSMOPOLITAN BODIES IN THE GLOBAL ARENA

The practices that both beauty pageant contestants in Nigeria and sex workers in Vietnam engage in are emblematic of their countries' respective shifts into the global economy. By examining the similarities between women's embodied practices from two distinct parts of the world and in two different industries, it becomes possible to examine how people in developing nations work to establish themselves as major players, not just in the cultural economy of beauty, but also in the broader global political economy. While Western standards of beauty have dominated the global market in past eras of globalization, the process of homogenization, though powerful, was never complete. The more recent era of globalization—since the 1980s—enables alternative visions of beauty (Jones 2011) that are much more nuanced, multidimensional, and engaged with local and regional ideals.

The rapid economic growth in Nigeria and Vietnam, alongside China and South Africa's growing prominence in the global economy, has created new openings for women in both nations to redefine and refashion the sites of attractiveness and raise awareness around new definitions of international ideals that emerge from local spaces and through regional relations. Pageant contestants and sex workers both undergo tremendous bodily transformations that signal not only their individual mobility, but also that of the nation. Pageant contestants bridge the gap between Nigeria and the rest of the world by using their newfound fame to help promote positive images of the country in a community of nations. Sex workers in Vietnam, on the other hand, play a critical role in helping elite businessmen strike business deals with foreign investors. Their surgical enhancements are often highlighted in these spaces to illustrate the dynamism in Vietnam's economy as it reaches even the poorest women from nearby villages.

The cultural economy of beauty is continually changing as global forces interact with local cultures. Together these women living in disparate parts of the global world are redefining and expanding cultural ideals of international standards by shifting the trends within their own countries that other nations may someday begin to emulate. This does not discount Western influence around the world; rather, we illustrate the complexity of cultural flows as women in local spaces react to and find ways of establishing new ideals. We can no longer think of Western ideals as equal to cosmopolitan whiteness (Saraswati 2010) or as the international standard that women around the world seek to emulate. Instead, these women are actively refashioning new ideals that combine global, regional, and local standards of beauty. This focus on shifting international standards is a framework that does not apply only to a beauty culture; rather, it is a framework that circulates in the arena of politics and the economy. As countries like Nigeria and Vietnam establish themselves on a global stage, they are legitimizing new ideals that are regionally and contextually specific. Consequently, women's bodies play a vital role in representing their nations as rising members of the global economy, constructing a space for the creation of new "international" standards that allows for the modification and reconstruction of embodied cosmopolitan femininities that can come to influence others around the world.

NOTES

1. For a more detailed description of our respective research methods, please see Balogun (2012) and Hoang (2011).
2. These earnings were very high in the local economy, as white-collar professionals with master's degrees in finance typically earned roughly $1,000–2,000 per month.
3. The state is divided into mainland and Island areas. Many places in the Island are known for their luxury estates and have a more suburban feel.

2

Aesthetic Labor, Racialization, and Aging in Tijuana's Cosmopolitan Sex Industry

Susanne Hofmann

Introduction

Tijuana's sex industry is an extremely professionalized and competitive sex work location where women engage in high levels of aesthetic labor. I will delineate different strategies of aesthetic labor that female sex workers deploy in order to maximize their financial gains in this trade. Aesthetic labor refers to the incorporation of workers' embodied attributes into the labor process (Tyler 2012); it involves the creation of a physical appearance that is suited to the work environment. An "aesthetically pleasing performance" is expected in most service industries, especially those that boast a predominantly female workforce (Wellington and Bryson 2001, 934). In the context of sex work, some scholars utilize a concept of aesthetic labor, which indicates a conflation of work on one's appearance with emotion work (Ditmore 2007; Sanders 2005a; Zheng 2009). Sex workers' compliance with the aesthetic nature of "the prostitute role" constitutes a *performance* that is part of a business strategy to attract a steady clientele (Sanders 2005a, 334–335). Women actively manipulate their appearance to fit the desired commodification of attractiveness, body parts, and sexual acts and thereby to "capitalize on sexuality" (Singer 1993, 39). In this context, it is important to highlight that the heteronormative aesthetic appearance[1] created for the work environment is not necessarily transferred by individuals into the private sphere. Rather, through aesthetic labor, women emphasize and effectively manipulate aspects of heteronormative femininity in order to capitalize on this financially.

This research was conducted in several months of fieldwork in 2006 and 2008. I interviewed a total of 25 sex workers, a majority of whom were interviewed at the Department of Health Control (*Departamento de Control Sanitario*), a governmental health center for sex workers located in the city's red light district. All sex workers interviewed were female and between 18 and 53 years old, and most of them were internal migrants originating from states other than Baja California. Furthermore, I interviewed 12 individuals who worked with sex workers in their professional lives, including health officials, psychologists, administrators, and NGO workers. Beyond that, I was able to carry out ethnographic observations in about 12 different sex establishments[2] (clubs, bars, and one gentlemen's club). In those venues, I talked informally to clients, sex workers, and staff members, but did not conduct formal interviews.

Most of Tijuana's sex workers are self-employed and work in one of the clubs or bars of the red light district, which is only a ten-minute taxi ride away from the border crossing point. There are different "classes" of establishments that target specific clienteles; high-priced, upscale venues seek to attract a more middle-class, professional clientele from both Mexico and abroad, whereas a greater majority of locations cater to a less affluent working-class clientele of Mexican origin. While upscale clubs provide a modern, glitzy interior, as well as entertainment, such as striptease, lap dancing, pole dancing, and foam showers, the lower-priced venues have a more basic décor and specialize in live music, partner dance, and providing clients with company for the evening.[3] The classed sex establishments noticeably employ a purposefully selected workforce, catering to different racialized desires and tastes. While upscale venues provide a workforce that is lighter-skinned, of statuesque beauty, and with the looks of models from an American fashion magazine, working-class establishments employ women who are generally shorter, darker, and some also older. With working conditions being better and money-making capacities much higher in upscale establishments, dark *mestiza*[4] and indigenous women are clearly disadvantaged. Inside the clubs, high-class sex workers make money from *variedades*, stripping on stage for a set of three songs, collecting tips from customers in their underpants, and from *privados*, private lap dances, which can be provided either at the table or in a specific booth. Depending on the degree of nudity, private dances cost from 20 to 60 dollars, from which the management takes a quarter or more. In working-class bars and clubs, women generate income through *fichas* (much more than they do in upscale clubs, even though the *ficha* or drink coupon system exists there as well) and through partner dances for a dollar per song. Women from both types of work sites sell sex in separate hotels that generally belong to the same owners, and are located either next door or opposite. All the money they make from selling sex is their own, and women generally negotiate prices and services themselves, yet clients must pay the management a fee per woman leaving the club. Overall, incomes vary considerably in relation to the prices of the *fichas* and the percentages of commission charged from lap dances. Some contracts, for instance, include that women pay a certain amount of *fichas* to cover the time they spent with clients selling sex. These are attempts of managements to directly tie their profits to the earnings of sex workers. Selling intercourse for between 60 and 120 dollars, young, attractive women generate incomes between 500 and 1000 dollars on weekend nights. In the red light district, one also finds a few gentlemen's clubs and massage parlors, however, most of them are dispersed throughout the city. Additionally, there are at least as many women selling sex in the streets as there are indoor workers. The social characteristics of street workers differed significantly, with many women being indigenous or from poor families, either with a background in agricultural labor or the urban informal labor sector; many possess little schooling. While most indoor workers are independent, many street sex workers work with pimps, and their earnings are considerably less.

BODIES ON THE MARKET: BEING "REALLY MADE UP!"

In Tijuana's extremely competitive sex industry, bodily appearance at work requires perfection. On weekends, clubs are packed with often over 200 women, generating very challenging working conditions for women who suffer not only from scarcity of clients, but also from hostile relations between workers, both of which many bemoaned when talking to me. Clients lament a pushy atmosphere as Hemmingson alludes, "Hustle is the key word—with so much competition, an Adelita girl [a woman working in a venue called Adelita Bar] does not wait for men to approach her...They immediately go into the sales pitch, asking: 'Sex-o' or 'Fucky/sucky?' or 'Do you want to go upstairs to a room?'" (2008, 41). Consequently, in order to be selected by clients, sex workers must invest high levels of aesthetic labor, especially in order to secure return clients and maximize financial gain. In the course of their professional lives they learn how to hide certain body parts and emphasize others with makeup, hairstyles, body paints or glitter, lingerie, and erotic accessories. They learn that their personal anatomy has strengths and weaknesses in relation to the preferred body shapes or types of the sex trade.

The art of *arreglarse* (fixing oneself up) has to be understood in a localized context (see also Casanova 2004; 2012). *Arreglarse* is very much part of everyday talk in Tijuana. Mothers might say *"Arréglate!"* (Fix yourself up!) to a daughter before leaving the house, or boys might gossip, *"ella no se arregla bien"* (she doesn't fix herself up enough) about certain girls in their class. On the one hand, *arreglarse* means being properly and nicely dressed; it can be smart but does not have to be formal. It means that one's clothing has no holes or marks, fits well, and is fashionable. But then there is another meaning attached to it, which relates more to hair, nails, and makeup, ideas related to femininity, orderliness, the shaving of hair (armpits, legs, bikini zone, women's beard if applicable), plucking eyebrows, and drawing them accurately with a pen in the color of one's hair. All this is related to what is considered feminine and proper in local terms.

It also means caring for one's body, not primarily understood in health terms here, but more in relation to a local understanding of femininity and beauty. Most sex workers wear makeup that is discreet, as opposed to extravagant, and above all, applied with meticulous accuracy (that is, fine lines). The most significant aspect of the meaning of *arreglarse* is the effort that must be seen to go into one's appearance. Long hair, for instance, is only considered *arreglada* if it is combed and arranged in some way, for example a bun, or parting, not just loosely hanging down. Most women also use gel to straighten their hair, or to model some kind of hairstyle with plaits. The considerable effort that has been put into rearranging the natural needs to be noticeable. *Arreglada* means that the natural, such as "wildly" growing hair, for instance, has been brought under control. An ever-increasing range of practices and beauty products help to demonstrate one's effort.

Arreglar (the non-reflexive form of the same verb) also has the meaning of changing and mending, remodeling, patching, correcting, adjusting, and repairing; there is also a sense of being flawless, untainted. Damaged skin, for instance, has to be mended, and bitten nails hidden and treated by manicure. I would suggest that there is a particularly strong need to look flawless and untainted among disadvantaged subjects, such as sex workers at the border. Meredith Reitman's (2006) study on whiteness and fashion in the IT sector draws attention to the fact that people with low social status cannot risk not to accentuate their "arreglada-ness," since understating arreglada-ness is already a sign of status and privilege. Members of the middle classes, who possess and embody social, cultural, and economic capital,[5] do not need to be concerned about their appearance as much or in the same way as people without access to the same resources. Tijuana's sex workers' *arreglada*-ness has a slightly formal touch to it. *Arreglada* is not casual and certainly never exaggerated or gaudy. It is characterized foremost by meticulousness, accuracy, attention to detail, and complexity in the treatment of hair, nails, and makeup. It is a very complicated art and is therefore often performed by professional beauticians or stylists who operate inside the red light district. In the context of the sex industry, *ser arreglada* (being fixed up) is also linked to having a neat and clean appearance that suggests body hygiene and safety to the potential customer.

In Tijuana's sex business, one must have very fashionable clothes in order to keep up with others and be equipped for this competitive environment. Rosalía,[6] a 29-year-old indigenous woman and street sex worker who originates from the southern Mexican state of Puebla, reflects on her sparse income: "Maybe it is because clothes go out of fashion and they [the clients] only take women with brand-new, fashionable clothes. That's what is going on here." The income Rosalía makes from street sex work does not always enable her to buy new, up-to-date, and fashionable clothes. She already struggles with the obligatory payments for the required health tests,[7] her costs of living in Tijuana, the remittances to her family in the south, and travel costs for visits home, which she tries to make every two years.

Being *arreglada* or "fixed up" is of paramount importance when working in Tijuana's red light district. In order to compete with others, women need to maximize their bodily capital (Wacquant 1995; 2001; 2004). Beyond a meticulous maintenance of the bodily appearance, *arreglarse* also involves training one's body. In order to gain a professional advantage many sex workers purchase professional beauty services, such as makeup application, manicure, pedicure, eyebrow shaping or removal, and waxing, or gym memberships, costs that are perceived as sensible investments in one's bodily capital and which they hope will eventually pay out. In Tijuana's upscale or higher-priced sex establishments, high-standard aesthetic labor, or "*arreglada-ness*" is time and again turned into an instrument of repression in the workplace. Managers in these sites put pressure on women with regard to their bodily appearance. Paulina (29, from Durango) recounts, "Every two months they always cut the personnel and say 'We don't want you anymore because you don't make yourself up [*no te arreglas*].' Forget it, you have to be *really fixed up [muy arreglada]*! You are pressured all the time." Managers give heavier women a hard time, threatening them with dismissal if they do not lose weight. In order to comply with the market demands for novelty and variety, managers regularly lay women off, in order to exchange them for newcomers in the business. Since not being *arreglada* enough is immeasurable and hence subjectively attributed, it is an easy and frequently used tool that allows managers to restock their workforce and supply clients with a new variety of women.

PERFORMING FETISH PERSONAS

At higher-priced establishments, Tijuana's sex workers perform fetish personas, such as a leather rocker diva, a cat woman, or a 1970s hippie girl. The city's highly professionalized sex industry demands great flexibility and adaptability from women with regard to the personas they enact, as well as the associated financial costs. Some sex workers I talked to complained about the immense costs of the costumes and accessories for the ever-changing theme nights that venues put on in order to attract clients:

> Tomorrow we will have to wear costumes. On Friday there will be an event and we have to go as rockers. I will have to go and buy leather clothes, get my nails done, my hands, you can't work in any other way here.... Everyone has to wear leather and chains. Now you see me like this, but in there they comb us, we have to have combed hair, we have to put make-up on and shoes. We can't just come in any shoe; we have to wear ballerina shoes.... There are two people who comb your hair and do your make-up. Of course you have to pay them. They charge between twenty and fifty dollars. Imagine, for hair styling alone they charge twenty dollars!... All the women use this service. I have my own hair iron and hair dryer, my own comb and everything. But if you want something special, you go to them and they do it.... About twenty days ago there was an event with a "seventies and eighties" theme. We had to buy clothes, wigs and all the rest of it.

A worker who earns less due to her age or because her stature, body shape, or skin color differs from currently desired beauty norms, cannot keep up with these immense investments, further diminishing her opportunities to make money in this competitive environment of erotic styles, since clients go for women with the fanciest costumes and the most stylish appearance. The clients' choice works according to patterns of consumption: one wants to get the most for one's money. Sex

workers in fancy dress serve as objects of prestige for clients
and also reveal a competitiveness among clients. Only the most
affluent customer is able to purchase the services of the best-
looking and most creatively made-up woman. Thus, the fancy
dress parties also disclose a pecking order among clients.

To a degree, the fancy dress codes can also have the potential
to level existing inequalities in natural bodily capital, since an
additional element such as inventiveness or creativity comes into
play. The red light district has given rise to a whole new market that addresses this need for erotic accessories. Shops selling
erotic fashion and accessories such as a plush thong with devil's
tail, for instance, or a G-string with cat tail, are located in Calle
Coahuila, in the heart of the Zona Norte (Tijuana's red light
district), and also in several hotels. Most women do not have the
time or means to make such clothes more cheaply themselves.
The multiplicity of rapidly changing themes makes it impossible
to keep up producing such accessories independently. Sex worker
Leonora, for instance, has found her own niche in this special-
ist market of erotic fashion. She has established her own sewing
workshop in her house to which she invites her colleagues to
pop in and be measured for the clothes she makes them.

The competitive environment of sexual entertainment
requires a great deal of aesthetic labor, creativity, and entrepreneurial spirit, and the diversity demanded by the industry
requires sex workers to spend significant sums on their appearance. One of the above-mentioned cat or devil costumes, which
consist of hardly any material, can easily cost 30 dollars in one
of the local shops, the equivalent of a medium-length lap dance
in one of the better-paid clubs. Sex worker Paulina told me that
women attempt to resist theme nights by staying away, which
can be countered by harsh penalties: "They [the managers] are
very demanding and if you don't go that day, they punish you
fifteen days [they are banned from work for fifteen days]."

"EVERYBODY WANTS BOOBIES": COSMETIC
SURGERY AND SEX WORK IN TIJUANA

One way in which sex workers respond to the ever-intensifying
competition of bodies in Tijuana's sex trade is by obtaining

cosmetic surgery. The pressures of fashion and beauty standards are immense within the sex industry, and most women have thought about their personal stance with respect to cosmetic surgery. Within reach of Tijuana's city center there are numerous clinics that perform cosmetic surgery. This abundance of cosmetic surgery clinics available in Tijuana is related to the border city's economic interconnectedness with its rich and powerful neighbor the United States, whereby service provision has historically been directed at the needs of American citizens.[8] Sex workers who work in upscale bars can afford to pay the three to five thousand dollars for an operation after a relatively short period of saving. Hence, many women ponder the advantages and disadvantages of cosmetic surgery, and decide for themselves which kinds of body work, practices, and adaptations are acceptable for them. Paulina highlights the popularity of cosmetic surgery in Tijuana's sex industry: "Most women in my bar have had an operation. Not me, though. But in general, everybody wants boobies [she uses the English term]."

The occupational experience of sex workers is characterized by an engagement in "corporeal self-production" (Wacquant 2001, 188). In their professional practice, sex workers reshape their bodies in accordance with the body types most popular with their male clientele. The bodily work and practices women engage in aim at producing a specific type of body that achieves the highest possible income. At present, the performance of heterosexualized bodies[9] appears to be the most income-promising body type. By capitalizing on heterosexualized bodies, women appeal to their male clientele and boost their incomes. In transforming their bodies they "literally produce a new embodied being out of the old" (Wacquant 2001, 188). The most common surgical changes that are obtained by Tijuana's sex workers are changes of noses, lips, hips, buttocks, and breasts. Paulina reckons:

> Women who have had cosmetic surgery make more money, because it is in vogue. Three, four years will pass and then it will be in fashion that they [the clients] like them fat and then six years later they like them thin. There is a time for everything. It's human nature, isn't it? Every other minute, we change our appearance. But if you get more money, sure.

Strategically modeled bodies are the way forward in Tijuana's sex industry, securing women increased incomes, more regular clients, and access to higher-priced establishments. Paulina considers cosmetic surgery a present fashion that will eventually pass. She has not had cosmetic surgery herself, but acknowledges the practice as an intelligent and strategic move to generate more income more quickly. Despite the popularity of the remodeled body, another sex worker, Serena, believes that the natural body still possesses its own unique and positive charm, and that there is more to sex work than just bodily appearance:

> There are lots of young girls with well-made bodies. And when I say 'well-made,' I mean in the sense of surgery. They have had surgery and don't have their normal bodies any longer. This has a lot to do with the fact that everybody likes an artificial physique. However, many [clients] do like a natural physique and sometimes they prefer good treatment, someone to talk to or to trust—someone who listens. The prostitute-client contract is not so simple.

From Serena's narrative we learn firstly, that clients come to sex workers with a multiplicity of tastes, desires, and needs, and secondly, that women make their own decisions as to which aspects of their appearance or performance they want to modify. Some choose to invest in appearance, figure, or adaptation of their bodies, while others decide to develop their emotional, affective, and communicative skills. The latter constitute a market niche for which considerable demand exists.[10]

The particular aesthetics and visibility of cosmetic surgery require a liberal culture that allows women to live comfortably after deciding to undergo cosmetic surgery. In order to live with significant enhancements of breasts, buttocks, and lips, one needs an affirmative environment tolerant of body modifications that not all the towns and villages that sex workers return to offer. Mexico's metropolitan centers, where cosmetic surgery has reached a state of ordinariness due to the influence of media, advertising, and *telenovelas*,[11] are much more likely to provide such tolerance. Elisabeth, who moved to Tijuana from a small peasant village in the southern state of Puebla, feels that since living in the border city she has already adapted her appearance significantly. In her personal life outside sex work, she wears neat clothes of North American style: blue jeans, youthful shirts, and white sneakers. While in the workplace she dresses according to the exigencies of the sex industry. Surgical body modification, however, is not a possibility for her, as she describes:

> I am a normal person [común y corriente] as you have seen. I have improved my physical appearance. I have not done anything to myself here [pointing to her breasts, bum/hips, face]. It is not because it wouldn't be pretty to add something or change parts of the body, but for me my children are the most important thing. And if I change my physique, they will ask me why: "What happened to you? Why are you walking like that?" [laughs] Well, no, better not. I respect my family. I love my children and it's for their sake that I want to get ahead in life.

Elisabeth's thoughts on the possibility and realization of cosmetic surgery are embedded in her responsibility and commitment to her family. Sex work is a temporary option she has chosen out of necessity and not a lifestyle to which she aspires. However, Elisabeth does not judge her colleagues' practices of cosmetic surgery. She knows that it is not an option for herself, but demonstrates a high level of tolerance and acceptance. The irreversibility of the decision to obtain cosmetic surgery requires the appropriate social background. Attitudes toward cosmetic surgery are also partly age related. The younger a person, the more likely it is that her social circles will accept her surgical body modification, due to processes of social change. Anonymous urban contexts make the choice of cosmetic surgery more feasible. In her peasant village, however, Elisabeth would stick out with a modified body. It would be obvious that she is a sex worker, and her profession would thus be revealed to her entire social circle. In making this disclosure, she would expose her family to ridicule and contempt, which she wants to prevent, in order to protect them from the stigma that could significantly hinder the social progress she wants for her family.

Despite widespread acceptance of cosmetic surgery among Tijuana's sex workers, the topic frequently causes disputes. Leticia explains:

> When you don't have anything operated on, the ones who are operated fuck you over more. You wonder why, if you haven't done her any harm? She stands out because she is operated on and gets a lot of work. But the one who is not operated, she is envied because sometimes the client wants her precisely because she is natural, spontaneous, and has her own mind. And this causes envy and anger among women who have been operated on.

Women who have had surgery are in high demand in Tijuana's sex industry, and their modified bodies generally allow them higher incomes. Leticia's remarks suggest that the income differentiation between operated and non-operated women does create tensions among sex workers. However, her comments also highlight the limits of marketability of surgically modified bodies. Women who have been operated on can no longer represent the "natural type" for those clients who desire unmodified bodies. The provision of one body type only does not respond to the broad range of tastes that must be catered to by sex workers in the sex industry; multiple body types are viable in the professional sex industry.

MIMING WHITENESS/PERFORMING SAMENESS: RACIALIZED SEXUALIZATION IN TIJUANA'S SEX TRADE

The attentive observer—say, a visitor or client in Tijuana's red light district—will quickly realize that there are significant differences between the street and indoor market, as well as among different classes of clubs or bars. Among the taxi dance halls, massage parlors, gentlemen's clubs, strip clubs, bars, and cantinas, there is a tendency that the higher the standard of the establishment, the whiter (in terms of skin color) the selection of women. The contemporary racialized hierarchy of Tijuana's sex industry enables white, blonde, and blue- or green-eyed women to charge the highest prices.

With whiteness providing access to better employment conditions and higher incomes in Tijuana's sex industry's racialized hierarchy, sex workers make efforts to fulfill the corporeal profile that is in demand on the market. Many sex workers in Tijuana's sex industry have hence dyed their hair blonde and many also wear blue or green contact lenses. Shaping one's corporeal appearance toward what can be considered a mainstream ideal of beauty in the "West" is popular in Tijuana's red light district. Following Bhabha's notion of "colonial mimicry" (1984, 125) I understand women's body modifications not as mere adaptations, but as strategic appropriations that allow them to capitalize on racialized desire. Women selling sex in the Mexican border sex industry can be understood as skilled mimetic practitioners in terms of adapting "foreign" corporeal styles as well as "foreign" communicative and cultural styles (Hofmann 2010, 240). Sex workers invest money and energy in producing a corporeal appearance that blurs the difference between what can be recognized as "familiar" and "foreign" bodies by foreign clients. In addition to miming whiteness by deploying procedures such as the bleaching of hair and wearing of blue or green contact lenses, part of the aesthetic labor that sex workers in Tijuana perform is adopting foreign cultural styles and performing cultural sameness, often signaled in clothes or accessories that display "Americanness." We can understand Tijuana's sex workers' strategies better if we look at excerpts from sex tourist websites.

Clients who exchange their experiences of purchasing sex in Tijuana's red light district in online forums, such as ClubHombre.com, emphasize that they prefer tall, white, and thin women rather than small and dark women, or women with curves, as becomes clear from sex tourist Brockton O'Toole's (2013) comment:

> The vast majority of Chicago Club girls are taller, thinner, better dressed, better made up and simply better looking than the other women in TJ [Tijuana]. Of course, beauty is in the eye of the beholder. When I make this observation, I do so knowing my taste runs toward the North American, European, MTV ideal of female beauty. If your tastes are different, YMMV [your mileage may vary].

Statements confirming that clients who consume sexual services across the US-Mexico border look for something that
is "the same" or "the familiar," rather than "the foreign" or
"the exotic," are common in such user forums. Various studies on sexual encounters in tourist locations have pointed out
that clients who solicit sexual services abroad, in different cultural contexts, tend to exoticize women (Cabezas 2004; Enloe
2000; Kempadoo 1999; Ryan and Hall 2001; Wonders and
Michalowski 2001). Sex tourists make particular associations
between nationality, race, and sexual prowess. Foreign, nonwhite women are imagined in idealized and exoticized ways as
softer, more loving, more caring, and more female than white
women, as well as "naturally" more willing to engage in sexual
activities. The above-mentioned literature suggests that othering, the process of eroticizing and exoticizing nonwhite women,
is something that generally takes place in intercultural commercial sexual exchanges in countries of the global South. In
the context of the sex industry in Tijuana, however, it seems
that male clients do not seek so much the "exotic" Other, but
rather look for the familiar and known body that is in compliance with beauty standards that have emerged from the global
North with respect to height, body shape, and skin color (Black
2004; Eco 2004; Gilman 1999; Gimlin 2002; Jeffreys 2005;
Kuczynski 2007; Tate 2009). US clients—treating Tijuana as
their own backyard—look for the familiar and known, just at a
cheaper price. Existing economic inequalities are integrated into
a matrix of desire together with differences across race, ethnicity,
and nationality at the border. In the US-Mexico border context,
we therefore find a particular kind of othering that does not
necessarily demand a nonwhite body, but a body that is available
at a minimal price.

   In many studies on commercial sex in the global South,
racialization and exoticization of women of color by white
sex tourists have been central to the analysis (Brennan 2004;
Cabezas 2009; Kempadoo 1999; O'Connell Davidson and
Sanchez Taylor 1999). Kempadoo's volume *Sun, Sex, and Gold:
Tourism and Sex Work in the Caribbean* (1999) was one of the
first works to address racialization in the context of sex work.
The volume's studies highlight that in tourism-oriented sex

work, racialized and ethnic differences are critical. Clients are foreign by culture, language, and often race to the sex worker, with the "Otherness" of the sex workers being a source of desire for the clients (Kempadoo 1999, 21). However, most of these studies deal with a situation where mostly white tourists purchase sexual services from women of color. Studying the sex industry in the Mexican border city Tijuana, I realized that these racialized sexualizations play out somewhat differently there. Tijuana's red light district is a truly cosmopolitan place where the effects of globalization become visible. Transnational and transcultural encounters take place every day in Tijuana's vibrant sex trade. Clients who enter they city's red light district come from a broad range of national, ethnic, and racial backgrounds, and hence we do not find a clear division between sex providers as women of color and the buyers of sexual services as white, but in addition to white clients we find individuals such as black servicemen, Lebanese businessmen, local Mexican-Chinese, Japanese, Koreans, Mexican-Americans, and (white and *mestizo*) Mexicans.

In her ethnography *What's Love Got to Do with It?* (2004), Brennan develops the concept of "sexscapes" as sex trade locations that are embedded in international travel routes of individuals from the global North to the global South. A distinguishing factor for sexscapes is that in those locations "the sex trade becomes a focal point of a place and the social and economic relations of that place are filtered through the...selling of sex to foreigners" (Brennan 2004, 16). According to Brennan, sexscapes in the global South are sites where sex tourists can not only buy sex more cheaply than in their home countries but also live out their racialized fantasies. In fact, Brennan claims that associations between nationality, race, and sexual prowess draw sex tourists to sexscapes in the global South. The economics of sexscapes depend on racial differences between the buyers (sex tourists) and the sellers (sex workers) (Brennan 2004, 32).

However, in Tijuana's cosmopolitan sex industry, we do not encounter such a strict racial division of sex tourists (white) and sex workers (women of color) as Brennan found in the Dominican beach town Sosúa. I therefore argue that the racialized sexualization we find in Tijuana's cosmopolitan sex industry resembles more Agathangelou's (2004) study conducted in the Turkish-Greek context: in Tijuana, light-skinned women constitute the highest-priced "object" of desire and are located at the top of the sex work hierarchy. Brown bodies are devalued by many clients, and the indigenous female body is considered the least desirable, thus located at the bottom of the income ladder. Consequently, many women have erased "exotic" or nonwhite aspects of their bodies, and attempted to "whiten" their appearance (for example with contact lenses, hair color) as much as possible. Light-skinned women work in better conditions and for higher pay than darker-skinned women, a fact which reveals that the sex industry reproduces colonialist racist hierarchies. Agathangelou (2004) has explained these dynamics with the desire of brown men to "whiten" themselves by privileging white bodies over brown ones in commercial sexual encounters. Light-skinned sex workers become a fetish that allows brown men to increase their value within a racialized hierarchy of masculinities. Underlying Agathangelou's concept of the "whitening" of men of color by selecting white women to satisfy their pleasures, is the idea that one's "measure of whiteness" (Pinho 2009) is not only defined by skin color, but also by gender and class affiliation. Both masculinity and class status of "not quite white" men are unstable. By consuming white bodies rather than brown bodies in the context of the sex industry, "not quite white men" satisfy their desire to be "white," thereby "silenc[ing the] social conflicts and contradictions such as his incomplete access to 'whiteness'" (Agathangelou 2004, 157). White and light-skinned sex workers in this context have the status of a fetish, decorating the "not quite white" bodies of the men who select them for sexual services. In the case of Tijuana's sex industry, we also find that differences of race, gender, class, and nationality become eroticized and commodified inequalities whereby buyers eroticize differences and sellers capitalize on these differences, yet unlike commercial sex in many other locations of Latin America and the Caribbean, for Tijuana's sex workers the ability to perform whiteness is the most treasured asset.

The othering of women in the sex trade takes a particular shape in Tijuana's sex industry. The racialized sexualizations that operate in Tijuana's sex trade resemble what O'Connell Davidson and Sanchez Taylor (1999, 46) distinguish as "denigrating racism" in opposition to the "exoticizing racism" that many studies on sex tourism locations have found. Neither when talking to sex tourists in person, nor when browsing through experience reports of sex tourists who have visited Tijuana online, did I find much explicit reference to race or ethnicity. However, commentators on the sex tourist website worldsexguide.com reveal through their indications of certain streets and venues they frequented, that they must refer to indigenous women with particular negative comments, since they mention sites in which indigenous women work in great numbers. For instance, talking about buying sex in the streets of Tijuana's red light district, the forum participant Smartdummy posted: "Remember, though, these girls have low, low prices, but make it up in volume"—translated—"don't expect anything tight," and he continues, "the women think they are the best thing since flat tortillas." Another anonymous sex tourist makes an even more blunt racist comment devaluing the bodies of indigenous women, by saying: "Skin tone ranges from African semi-dark to white, but the vast majority is ethnic mix or Hispanic, but not with Indian features. The Hispanics are Latina cuties, and do NOT look like Mexican 'Indians' with stocky frames or flat faces." As these examples from the online forum worldsexguide.com have shown, the racialized fantasies and desires of clients in this particular border city positioned dark, brown, or indigenous-looking bodies as undesirable and of lesser value, whereas a white "MTV ideal of female beauty" is considered desirable. This is not to say that exoticizing racism does not occur in Tijuana's sex trade, yet accounts of sex tourists in Tijuana reveal that frequent experiences of robbery and the undisguised aversion they encounter in the red light district does not leave much imaginative space for the kind of exoticizations found in beach sexscapes of the global South (Hemmingson 2008; Warren 1990). Additionally, many of Tijuana's sex workers make little effort to disguise the commercial nature of the interaction,[12] leaving little scope for exoticizing imaginations of them as warm, affectionate, and caring nonwhite women.

Kamala Kempadoo's (1999) edited volume includes studies on sex work in the Caribbean. The studies discussed in this volume frame contemporary sex work in the region as a legacy of colonial power relations. In the context of international and racialized relations of power, sex work can also be viewed as a resource for the sustenance and nurturance of the first world, which supports the refashioning of Western constructions of gender and sexuality, or as a haven for the nurturing of Western bodies and productive labor (Kempadoo 1999, 27). Sex work thus is one strategy of everyday struggle by the disadvantaged to tackle the unequal distribution of resources that exist on a global scale, and generate some gains and prospects for themselves.

Commercial sex industries operate in accordance with the contemporary neoliberal world order, which is racialized and gendered (Agathangelou 2004). By demanding the continuous provision of sexual services, provided by the broadest possible range of socially and economically disadvantaged women, the border-crossing client becomes an accomplice of the existing neoliberal world order. The clients in Tijuana rely on the asymmetrical relations that are produced across the border, and their desire rests on the (re)production of raced and classed masculinity and femininity in both bordering nations (Agathangelou 2004, 155). Clients presume Mexican sex workers to be available at a cheaper price or to provide more services for the same price, taking for granted and benefitting from the economic differences between the two border economies. The cultural encounter that takes place at Tijuana's sex market between disadvantaged women from the global South and cosmopolitan men belonging to a transnational capitalist class (Sklair 2001) confronts both parties with internalized stereotypes of cultural otherness and the sexualizations and racializations that are connected to it. We can understand the strategies of miming whiteness and performing sameness not only as a response to a particular market demand, but also as a part of a self-creation process that aspires to upward social mobility. Furthermore, attempts to appear white can also be considered a strategy of self-protection in a region, such as the US-Mexico borderlands, in which racism and resistance to immigration have flourished (see, for instance, Akers Chacón 2006; Benton-Cohen 2009; Dunn 2009; Fernandes 2007; Koskela 2010; Nevins 2002).

Aesthetic Labor against the Clock: Body, Time, and Sex Work

Many sex workers I spoke to emphasized that men prefer young women and fresh faces: women who are new to Tijuana and inexperienced in the sex trade. The bodily capital acquired in the course of sex workers' occupational activities is not inalienable personal property, gained once and for all. Sex workers are very aware that the bodily capital that they have acquired and fostered is finite. The signs of an aging body cannot be hidden entirely, and sex workers are aware of the short time frame available to them for economic gain in their profession, as Serena's account demonstrates:

> Well, look, unfortunately all of us start to work and we say "I will work, I will save money for this," but then there's some family situation, this or that happens, time passes by, suddenly there comes the moment when you realize that time has been flying and you have become old. The problem is, when we get old we have no security for our old age. So, when you have got old and haven't saved up money, this is when your situation is fucked up [*ahí se jodió la cosa*]. Who will look after you?

Several sex workers I spoke to told me that they were unable to stick to their original time schedule. Many had come with the idea of selling sex for approximately five years, however, unforeseen circumstances of different kinds prolonged their stay in Tijuana's sex industry. When sex workers are old, they have neither social security nor pensions: they have to pay for their medical costs, despite the declining incomes that older sex workers make in the business. Women who realize that they have become too old for sex work, and that their bodies no longer enable them to make a profit, are in a very difficult situation if they have not saved up for their future or invested in something concrete such as a house, business, or education. Serena thinks they should be given the opportunity to have social security, health insurance, and pensions like any other worker. We discussed German laws on sex work, whereby all registered and documented workers pay income tax and make contributions to social security systems providing health care and a state pension. Serena prefers this model, where women pay social security contributions and can then rely on the system in their old age. She said, "You know, nothing in this life is forever. The only thing that's for sure is death." She advocates for social security payments, even for women who work with pimps, because these will dump them once they are older.

Aging is problematic for sex workers, not only because they might be made aware of their age by clients or managers, but because it increases their vulnerability in various ways. Nevertheless, many have to continue selling sex until they have found other options. Older street sex workers are frequently physically threatened and beaten by younger sex workers, who fear that the older women's presence might diminish their own prospects for gaining affluent clients, who may be put off from strolling further into an area where older and less attractive women are situated (Bautista López and Conde Rodríguez 2006; Hofmann 2005; 2010).

Tijuana's sex industry offers a market for the services of middle-aged and older women who sell sex mostly to lower-income, older clients. However, working conditions for aged sex workers are grim, since their income ranks at the very bottom, and they are forced to take every opportunity that comes along. Sometimes hardship pushes them to take clients who do not want to use condoms. In relation to the incomes of older women, condoms are very expensive, and many can no longer afford them, thus risking their health. Due to their reduced income, older women can sometimes not afford to keep up with the obligatory health tests either. When caught working with expired work permits, they often experience police harassment, and face prison sentences of up to 36 hours. Furthermore, older women are more vulnerable because they are less capable of defending themselves in physical confrontations with malicious clients.

Time is money—this saying applies in a very literal sense to sex work. In order to counter declining incomes that come along with aging in the sex trade, women attempt to make as much money as quickly as possible, making rational time management one of the most crucial skills of the trade. Effective time management is an essential requirement of successful entrepreneurial activity. Sex workers need to not only calculate time rationally in order to utilize their bodily capacities in the most effective way on an everyday basis, but also take aging into account when planning their work trajectory. Sex workers have explained to me how time plays a decisive role in relation to the kind of sexual services they provide. Débora explains notions of "working slow" and "working fast" in Tijuana's sex industry. When she first mentioned that some women worked fast in her club, I asked her whether she meant that they work more hours than those who worked slow. She laughed at me and said "Oh, come on, you know what I mean!" Contrary to her assumption, I had no idea what she was referring to. Débora elucidates:

> Women who work fast do *privados* [normally lap dances in private booths] and they do other things that they are not meant to do there. Well, I don't know what they do [laughs] but you understand me.

Débora's club is located on Tijuana's tourist mile, the Avenida Revolución, and is licensed as a strip club only. This means women are not supposed to sell sexual services other than dancing on stage and dancing in private cabins for individual customers. Obviously, it is difficult to control what actually happens inside the booths, and women take advantage of the clients they have acquired, selling them blow jobs, hand jobs, and sexual intercourse despite not being supposed to, since they are officially registered as dancers only. "Working slow," on the other hand, means drinking with customers (*fichar*), performing strip acts on stage, and doing private lap dances for customers. Débora, for instance, sells dances for 40 and 60 dollars, of which she earns 35 and 45 dollars respectively after paying the fee to the club. For stripping there is no fixed price as such. Performers do sets of three songs, depending on how many customers are present, like their dances, feel attracted to them, and are generous. They earn however many dollar bills customers put into their underwear (minus a fee for the *talachero*, a helper who clears the stage, which is usually five dollars). An entire dance set might bring up to approximately 30 dollars for exceptionally pretty and skilled dancers. Compared to what can be earned by sexual services such as oral sex (approximately 50 dollars), vaginal sex (60–120 dollars), oral and vaginal sex (80–140 dollars), and anal sex (100–160 dollars), stripping and lap dances seem rather unprofitable. Débora explains that she is not in a hurry to make money and prefers to work slowly. Dancing being her sole activity, she accumulates considerably less money per night than her colleagues who do "unlicensed" *privados.*

Fast and slow sex work correspond to more or less effective forms of "corporeal entrepreneurship" (Wacquant 2001, 191). Sex workers measure the effectiveness of their work by calculating the effort invested against income generated over a certain period of time. Débora's explanation of fast and slow sex work demonstrates the significance of the temporal dimension in sex work. Rationalizing time[13] and enforcing time limits in sexual transactions are primary skills of professional sex workers.[14] "Working fast" means making as much money as possible within the shortest amount of time. While fast sex work allows the quick accumulation of money, it is at the same time more emotionally demanding, operates at a shorter physical distance from the client, and requires more bodily engagement. For the most part, strippers do not have intimate bodily contact with clients. While they often allow clients to touch them, intimate body parts are generally excluded. An awareness of one's declining bodily capital as the body ages, together with a need for the instant availability of cash, are the main reasons for women working fast in the sex industry. Rationalizing time is a constituent element of women's purposefulness and focus on their objectives. It is also considered to be part of professionalism in sex work. Women plan their time in sex work strategically, always with the intention of making as much money as quickly as possible, and leaving the sex industry as soon as their circumstances allow.

CONCLUSION

Aesthetic labor in the context of sex work means adopting an appearance that conforms to the aesthetic nature of the prostitute role. The professional performance of the prostitute role constitutes an exact and calculated method that utilizes the expectations of male sexual desire to make maximum financial gain in the minimum amount of time. Aesthetic labor forms part of a business strategy to attract customers and ultimately a steady clientele. I have described Tijuana's sex industry as a highly professionalized and competitive sex work location where women engage in tremendous levels of aesthetic labor. Women told me that it was extremely important to be *arreglada* and that in order to compete with other women (some of whom purchase the services of professional stylists), sex workers need to make sure to be "really fixed up." Being really fixed up means maximizing one's corporeal capital, and it involves the meticulous application of makeup and hairstyling, as well as shaving and training one's body. In order to gain a professional advantage many sex workers purchase professional beauty services or gym memberships, costs that are perceived as sensible investments in one's corporeal capital which they hope will eventually pay out.

Aesthetic labor in the context of sex work in Tijuana involves women's personal financial and creative investment in enacting a series of fetish personas. Tijuana's sex workers perform fetish personas, such as the leather rocker diva, the cat woman, or a 1970s hippie girl. Tijuana's highly professionalized sex industry demands great flexibility and adaptability from women with regard to the personas they enact, as well as the associated financial costs. Some sex workers I talked to complained about the immense costs of the costumes and accessories for the ever-changing theme nights that venues put on in order to attract clients within a very competitive industry.

Part of adopting heterosexualized bodies in order to capitalize on attractiveness and appeal to a male clientele is cosmetic surgery. I have described that Tijuana, historically a site that served the interests of its rich neighbor, has developed a vast cosmetic surgery sector. Young sex workers who possess a relatively high corporeal capital and work in upscale establishments generally manage to save up enough money for a breast enhancement operation in a few months. The women I talked to confirmed that breast enhancements are a good investment in Tijuana's red light district. Those who invest in their bodily capital are able to improve their earning capacities significantly. However, cosmetic surgery is not for everyone who works in the sex industry. For sex workers who come from rural villages or towns, getting breast enhancements or any other visible cosmetic surgery would be equivalent to coming out to their families and friends as professional sex workers, which some do not want. Women who decide not to have cosmetic surgery told me that they capitalize on naturalness, talking to clients, emotional labor, and their creativity instead.

This chapter has discussed that the aesthetic appearance that sex workers in Tijuana capitalize on constitutes of miming whiteness and performing heterosexualized bodies. Because whiteness is located at the top of Tijuana's sex industry's racialized hierarchy, sex workers attempt to appear whiter by bleaching their hair and wearing blue or green contact lenses. Part of the aesthetic labor that sex workers in Tijuana perform is adopting foreign cultural styles and performing cultural sameness, often signaled in clothes or accessories that display the American flag.

I have highlighted that aging is quite problematic for sex workers, not only because they are made aware of their age by clients or managers, but also because it increases their vulnerability in different ways. Preventatively many women counter declining incomes that come along with aging in the sex trade, by making as much money as possible in the present. Effective management of time, finances, and personal emotional and bodily resources is essential to ensure a livelihood of economic security after the marketability of the female body in the sex trade.

Globalization, producing mobility and travel of population groups around the world, has enabled new kinds of encounters in the sex industry at the US-Mexico border. We no longer find a straight division between sex providers as women of color and buyers of sexual services as white. Instead, clients who enter Tijuana's red light district come from a broad range of national, ethnic, and racial backgrounds. Sex workers, most of them women of color, adjust their corporeal styles and beauty practices to the changing desires of their cosmopolitan clients. While the aesthetic labor of cosmetic surgery and the desired corporeal shapes are increasingly global, I found that Tijuana's sex industry differed from other sex tourism locations in Latin America and the Caribbean with regard to particular racialized sexualizations. Unlike other sexscapes in the global South, in which brown bodies are the principal target of male desire, the racialized fantasies and desires of clients in this particular border sex industry positioned dark, brown, or indigenous-looking bodies as undesirable and of minor value, whereas white women were the more valuable asset, allowing brown men to whiten themselves within the hierarchy of racialized masculinities. Consequently, light-skinned women work in better conditions and for higher pay than darker-skinned women. Hence, regardless of the new subject positions and sexual encounters that globalization has brought about, there is a continuation of the colonial, racial hierarchy of bodies that is being sustained and reproduced in the cosmopolitan encounters that Tijuana's sex industry allows.

NOTES

1. By "heteronormative appearance" I mean an appearance that attempts to indicate heteronormative gender identity and heteronormative sexual orientation. A heteronormative appearance attempts to suggest to the viewer that she or he is seeing a female with a female gender identity whose sexual identity is heterosexual, for instance. Adopting a heteronormative appearance involves adjusting one's appearance to one that is perceived attractive to a majority of heterosexual men. This is most commonly achieved through adopting a feminine appearance (by means of clothing, hairstyles, and makeup).

2. Most of those venues were directed at a heterosexual clientele, however, I visited three locations that had transgender and male performers and were frequented as well by gay men and heterosexual women.

3. This activity is called *fichar*, which is comparable to hostessing: women sit with clients, talk to them, and drink with them for *fichas* (drink coupons), which are split between the bar and the sex worker. The work of *ficheras* (hostesses) often resembles what we know as the "girlfriend experience," or a genuine, romantic encounter, involving flirting and the client purchasing little presents, such as candy and flowers from street hawkers passing by.

4. *Mestizas/Mestizos,* people of mixed European and indigenous heritage, became the majority of Mexico's population (Menchaca 2001). *Mestizo* elites of the nineteenth and twentieth centuries sought to consolidate their political power by constructing a unified national identity, which omitted the various African, Asian, and Middle Eastern groups who have historically made up the Mexican population (Poole 2011).

5. In resonance with Bourdieu's (1990) term "habitus," I understand that social, cultural, and economic resources produce individuals who in the course of their lives come to embody the particular lived experience they have had. The social, cultural, and material resources sediment into particular embodied expressions, such as articulation, language, knowledge, skills, and posture among many others, throughout a person's lifetime, marking them as members of particular societal groups.

6. Both personal names and names of sex establishments have been changed in order to guarantee the confidentiality of research participants.

7. The most frequently required test is a combined HIV and cervical smear test, compulsory every four weeks, which costs $57–$60. However, most of Tijuana's sex workers are seasonal workers who must renew their health card (or work permit) any time they return to work. A renewal plus HIV test and smear test then amounts to a total of $83.20 (all prices from 2008).

8. Cosmetic surgery tourism to Mexico has not yet been a subject that received great attention from scholars. There are a few studies, however that allude to the US-Mexico border as health tourism destination (Herzog 2003; Muriá Tuñón 2010; Romero 2008), and to medical tourism in Mexico (Dalstroma 2012). For research into cosmetic surgery tourism to other Latin American regions, see Ackerman (2010), and on cosmetic surgery tourism more generally Casanova (2007); Bell et al. (2011).

9. By "heterosexualized bodies" I understand bodies fabricated to arouse heterosexual (in this case male) desire.

10. See also Bernstein's (2007) analysis of an increasing demand of "bounded authenticity" in what she calls late capitalist sex work.

11. A *telenovela* is a serial popular in Latin American television programming. The word combines "tele," which is short for *televisión* and "novela" meaning novel. *Telenovelas* are a distinct genre that differs from English language soap operas in having a predetermined duration and a concrete ending of the storyline. *Telenovelas* are aired in prime-time television schedule six days a

week and generally last about six months (between 180 and 200 episodes). Unlike US soap operas that tend to rely on the family as a central unit of the narrative, Latin American *telenovelas* focus on the relation between a romantic couple as the main motivator for plot development.

12. In his autoethnography on his experiences as client of sex workers in Tijuana's red light district, Michael Hemmingson describes, for instance, the aggressive advertisement strategies by street sex workers who "grab at the sleeves and arms of men walking by, hissing 'chhtt-chhtt' and quoting prices" (2008, 30). Furthermore, he frequently comments on the emotionally cold nature of the sexual transactions that take place in the hotels of the Zona Norte: "In the room, she would not look at me. She took her pants off and lay on the bed...There was more hate and loathing in her eyes than the one [another sex worker] last time" (2008, 81). This corresponds with statements from sex workers I talked to who claim that they strive for speedy transactions with clients of 20 minutes at most, charging extra for any excess time spent or for the fulfillment of special desires (which they also frequently deny). Hence, unlike some indoor sex workers who spent huge efforts into creating a pleasurable experience for the client, and keeping an "illusion of intimacy" (Lever and Dolnick 2010, 86) alive, many sex workers in Tijuana's sex industry focus on a quick turnover of as many clients as possible per night.

13. By "rationalizing" I mean applying a logic of efficiency. Rationalizing behaviors, in the context of work, strive to optimize labor processes by the means of time-saving measures, for instance. The ultimate aim of rationalization at work is to increase efficiency. Generally, measures of rationalization are imposed on workers by the management, however, self-employed workers, such as sex workers, might practice such behavior to raise their earnings and limit working hours.

14. O'Connell Davidson (1998) describes how "Desiree," an independent sex worker, guards against "time wasters" (clients who do not actually strive to buy sexual services, but come to ogle, check out, or shop around for sex workers) by charging an entrance fee to all visitors. Thereby, Desiree prevents them from having a thrill for free as well as wasting her precious time. Brewis and Linstead (2000, 89) discuss the importance of limiting time with the client as a strategy that "frames the sexual encounter." Measuring time is used as a device that enables sex workers to erect a psychological barrier between themselves and the client. In Sanders's (2005b)

study, sex workers used timing as a technique to control the sexual routine by spending the minimum amount of time earning the maximum amount of money. Brents and Hausbeck (2007) highlight that in the context of Nevada's legal brothels, the timing of sex work is now taken on and monitored by brothel managers. Bernstein alludes to changes in time management in late capitalist sex work, referring to high-priced services by indoor sex workers who refrain from strict time management, spending extended time with affluent clients, providing an experience of "bounded authenticity" (2007, 6). For further reading on the importance of controlling time in the client encounter see O'Neill (1996), and on an overview of sex work and time management, see Brewis and Linstead (1998).

3

IN PRAISE OF BIG NOSES

(*Personal Reflection*)

Persis M. Karim

When I was a child growing up in the suburban San Francisco Bay area town of Walnut Creek, I was aware of the ways in which my difference was cast in my look. Although my mother was from France, I had inherited many of my father's distinctly Iranian features: large brown eyes, thick dark hair, long eyelashes, plentiful eyebrows, and olive-tinted skin. Although my father was a handsome man, his exceedingly large nose and well-sized ears were an inescapable part of his distinctively foreign look. Until my adolescence I managed to avoid thinking of my nose at all. It was only when my older sister turned 19, and after a number of Iranian female relatives immigrated to the United States, that I became aware of how the bountifulness of my proboscis presented deficiencies to my blossoming beauty. When my father's two older daughters from his first marriage in Iran arrived in the United States, I was only ten, but I could see in their faces something of how they both resembled and didn't resemble my sister and me. Indeed, it was in that central region of their face, the nose, that they lost any familial connection. At one point in my early adolescence, I remember seeing one of my older female cousins at a family party for *Norouz*, the Persian New Year, with a large white bandage across her face. Although I was too shy to ask my cousin what had happened to cause the black and blue bruises under her eyes and the gauze bandages that concealed her nose, I surmised that it was part of an Iranian family tradition that might be visited upon me in the future. When I finally asked my father about my cousin's wounded face, he blurted out smiling, "well, my dear, in case you hadn't noticed, we've got a few people in this family with big noses."

The contagion of nose-reduction operations that I later came to know as "nose jobs" seemed more and more prevalent with each passing year. During the early 1970s, when Iran and the United States were close allies and when there was relative ease of movement between the two countries, plastic surgery had become big business in Iran. The influx of Western culture, US-made products, and the influence of American movies and television had transformed the landscape of aesthetics and beauty in that country. Iranian women sought to be more beautiful through blonde hair, smaller noses, and shorter skirt hems. My father, who arrived in the United States long before the largest number of Iranian immigrants came, was unaware of this trend; but when his female relatives began arriving in this country in the mid-1970s, they brought with them this same culture of beauty. Each year, when my Iranian relatives got together to celebrate the Persian New Year, I began to notice a tell-tale look among the women of my family: large brown eyes, voluptuous lips, and small, perfectly sculpted, Anglo-looking noses. Some were small and upturned, others were perfectly triangular and proportional; mostly though, they looked similar—as if they'd been reworked by the same plastic surgeon, who had taken away nature's unique character on each of their faces.

Around my fourteenth birthday, during that awkward phase when most adolescents begin exuding hormones and their features appear scarily large and unsettled on their still child-like faces and bodies, I realized I was being recruited into this cult of Iranian beauty. The first occasion was at my aunt's house when she told me that I should pluck my eyebrows and suggested a nose job so that I could become "more beautiful." On another occasion, my half-sister Lily gave me a lesson in upper lip depilation and gently suggested something similar. "You are beautiful, but we Karims have large calves and large noses." Even though I recognized the truth of her statement about the calves, I asked her what she meant about the noses. "You know, me and Cima (my other half-sister) both had nose jobs in Iran. You should get one too." Several months after that conversation, I became more aware that the odds were against me when I overheard my sister Avesta speaking quietly behind a closed door to my father about his financial support so she could get "the operation." I remember it being around the time of her twenty-third birthday. Although my sister didn't live at home, I knew that she had felt increasingly self-conscious about her large nose and its disproportionate size relative to the rest of her face. Unlike me, she possessed many of my mother's features: petite build, light brown curly hair, fair skin, and the recessive gene of green eyes. In our family, the nose job seemed to correspond with the approaching age of marriageability; the change of nose at early adulthood was in effect an indicator of being "on the market" for a suitor.

In the days after my sister's surgery, I saw her only once in the white bandages and with deep bruises under her eyes before she retreated to her apartment to recover and heal. After she recovered, she came over several times and showed off her nose like it was a brand new car. She held her head high and let her new nose, reduced and trimmed, signal her more ordinary and less ethnically marked face. For her, the operation had been a double success: she was more aesthetically aligned with American beauty norms and she had eradicated her Middle Eastern heritage, a part of herself with which she was never comfortable. She had also participated in a rite of passage that she shared with our two other sisters, and despite her distance from them, it was a way she belonged and I didn't. Of the four girls in my family, I was now the only one with my original nose.

I managed to avoid the topic of nose jobs until around the time I turned 18, when my father, who was acutely aware of the long-standing tradition and the rising costs of such an operation, asked me if I wanted to get one. He told me that along with savings he'd been putting aside for my college education, he'd also saved an additional $3,000 in the event that I wanted to join my sisters in the nose-reduction club. Although I appreciated the gesture, deep down, I was distressed by his offer. "But do you think I need one? Do you think I am not beautiful enough?" He laughed and responded by saying that I looked like him, and that as his daughter I would always be beautiful in his eyes. "But do you think I have such a big nose?" I pleaded. "Your nose is a smaller version of mine," he said. "I regard the nose as a sign of character, and on you it is part of the total aspect of your beauty. But what is inside is what determines how others see you." My father, who was deeply prone to philosophical answers, did not stop there. He answered me by offering an expression in Persian, "*cheshm-tan ashaq mibineh*"—it is your eyes that see beauty—quite similar to "beauty is in the eyes of the beholder," in English. "You are beautiful," he said, "but it is up to you to decide."

After thinking about it for several weeks, I decided not to get my nose "fixed" and instead took the money that my father put aside and used it for a trip to study abroad in my senior year of high school. I was aware that I was making a choice, an intellectual and feminist choice that would set me apart from some of my sisters and other female relatives. I was both proud and a little ambivalent about the choice I had made.

Ironically, it was years later that I would meet a man who possessed a large nose and who would later become my husband. Although he was Jewish and not Iranian, I felt a strange affinity for being with someone whose proboscis was even larger than mine. We both laughed at the idea that we came from cultures that sought to diminish the size of the nose, and that ultimately, we were better equipped to deal with the hotter climes with our large noses. Perhaps it was a subconscious appreciation for how my own father justified large noses in men with sexual virility and character. As I grew older I came to appreciate how my own nose belonged on my face, with my other features—large eyes, and a pronounced chin—but whatever the case, I no longer fretted about it.

Recently I watched the entire "Star Trek: Next Generation" series on DVD with my husband and nine-year-old son, encountering the race of people in outer space identified as the *Ferengi*. I became aware again how powerful the image of large noses is as a marker of foreignness and outsider-ness. (The *Ferengi* do not belong to the United Federation of Planets and are in fact a bit rogue.) In fact, the word "farangi" in Persian means literally "foreigner." The characters on this TV show are attributed with being greedy and solely motivated by profit—a characteristic that has been associated with anti-Semitic stereotypes, and more recently with both Arabs and Iranians. The negative association of the *Ferengi* on a popular late twentieth-century sci-fi television show was no longer about Jews but about other Middle Eastern people—Saudis, Iranians, and other oil-producing Arab nations that have become associated with avarice, profiteering, and a militaristic sensibility. In addition to being sinister and mercenary, the *Ferengi* are also unappealingly sexist; they are disdainful and disturbed to see the women of the Starship Enterprise serving alongside men in leadership roles.

More recently, I came to appreciate my large nose again when I learned in a 2009 news story on CBS by Jaime Holguin that Iran is now the "Nose Job Capital of the World." In a country where aesthetic beauty is limited by the efforts of the government to curb public displays of sexuality, both men and women seek to alter their appearance in response to the limited and restrictive codes of dress and bodily presentation. While liposuction and breast augmentation are the most prevalent procedures in the United States, in Iran, a nose job has become standard practice (at a mere $1,500). According to some news reports, the nose job is one of the few ways that women can and do alter their appearance when they are required to cover their hair and much of their bodies and are discouraged from wearing makeup. Ironically, this act of personal expression represents a rare opportunity for young Iranians who want to show resistance against anti-Western sentiments by the government and model images of European and American men and women in fashion magazines.

In 2006, director Mehrdad Oskouei captured the Iranian national obsession with nose jobs in Iran in a short documentary called, "Nose, Iranian Style" (whose title is a play on an earlier documentary called, "Divorce, Iranian Style"). In his slightly humorous documentary, Oskouei considers the epidemic of nose jobs that is estimated at 60,000 to 70,000 operations each year. The Italian photographer Fabio Bucciarelli documents the national obsession with rhinoplasty in his photo essay, "Nose Job: Iran, 2009," which appears both on his website as well as in the US-based online newsmagazine *Tehran Bureau*. His series of photographs, shot in Tehran in 2009, features young men and women who wear a single rectangular bandage across their nose in the postsurgical recovery stage. Each of his subjects (an equal number of men and women) pose for the photo as if to say, "I own my body, I own my nose. And I can alter my nose myself if I please." Unlike my sister, who hid her bandages and waited to reveal her "new" look to the world until after she was completely healed, this series of photos documents the seeming defiance encompassed by getting plastic surgery to look less Iranian and more Western. The bandage itself is seen as part of an aesthetic rebelliousness (see www.fabiobucciarellis. com/portfolio-item/nose-job/).

While I understand my own vexed relationship with my nose, I have also come to accept the body as a site for cultural, aesthetic, and personal meaning that changes over time and across cultures. I've come to accept my nose as both a sign of my heritage and have resisted altering it because of how others have construed its aesthetic value through a singular lens. Instead, I see my nose in a larger, global context that offers multiple significations and meanings.

At last, I turned to poetry to praise it:

In Praise of Big Noses

Persis M. Karim

I am the only one of four sisters
who hasn't gone under the knife.
I resisted the pleas of my aunt and sisters
to become "more beautiful," "more you."
I've kept my stately proboscis
intact—choosing not to excise its grandeur.

It suits me, I suppose—evidence of my father,
those people who live in the dryer, hotter climes
of the Mediterranean, in high desert plateaus,
cooling themselves with naso-thermo-regulation.
My old Jewish boyfriend used to say *how do the* goyim
breathe from those things anyway?

On my wedding day, my husband, also Jewish
and rather plentiful in that region of his face
completed his vows by saying "there is no guarantee in love,
but of this, I am certain: if we have a child he or she
will have a really big nose." When I nuzzle him
with mine, he pulls back his face, jumps

at the coldness of its tip. Contrary to popular belief, the nose
is not merely cosmetic—it can gauge temperature beyond the
    body.
And that's another thing, I've realized about the nose—
that smell is an underrated sense, perhaps a gift.
Imagine the possibilities for amplification: aromas
of jasmine, apple pie, saffron, lemon, rose,

might grow more intense, depending on the height
and angle of that fleshy mound. I admit to having no
scientific evidence for this, but I do wonder
what happens when a person alters
the things they were born with.

Whole industries were born from Iranian women
watching blonde, petite-nosed movie stars
who made them forget their own striking beauty
took thousands of years to evolve, only to be undone
by someone who decided that hairless, plucked, tucked,
sliced, nipped, and trimmed, were the loveliest

of them all. I like to think of the nose as great art
waiting to be discovered. Like those large-nosed kings
depicted on sides of temples, on papyrus, on caves, in
    colorful
Mayan pictographs like *Popul Voh*. Noses were signs
of nobility and prowess. Any king with a puny one
might have been thought of as small and impotent.

These days, I get a steady stream of emails offering penis
enlargement. But that's hidden, visible only
in bedroom interludes. The nose is the public display
of one's endowments—the relief map of a human face.
I study people's noses in order to read their origins—
to situate my gaze, to find how far out

in the world they really are.[1]

Figure 3.1   Proud, big-nosed Iranian American women (from l to r): Mona Kayhani, Persis M. Karim, and Aphrodite Desirée Navab.

Photo: Persis M. Karim

NOTE

1. "In Praise of Big Noses" by Persis Karim first appeared in the special "Iran issue" of *The Atlanta Review* 16(2) (2010): 45–46.

4

¡MÁS QUE UN BOCADO! (MORE THAN A MOUTHFUL)

COMPARING HOOTERS IN THE UNITED STATES AND COLOMBIA

Michelle Newton-Francis and
Salvador Vidal-Ortiz

INTRODUCTION

The year 2008 marked the twenty-fifth anniversary of the opening of the first Hooters restaurant in the United States, and the opening of one of four Hooters in Colombia. In the United States, the Hooters Girl is constructed as embodying the "all American cheerleader" and "surfer girl next door." These constructions connote a particular kind of femininity and sexuality, one that is decidedly white, US middle class, and heterosexual. Yet, this particular brand of female sex appeal cannot be exported or normalized around the world. Constructions of beauty are geopolitical; even with the impact of globalization, there are specific elements of racial phenotype, body shape, color, and curves that cannot be globalized. Indeed, popular cultural constructions such as the Hooters Girl and cultural spaces such as Hooters restaurants do not work the same everywhere, causing a tension between the activation of local adjustments in the face of external commercial business models. This tension prompts us to ask, how does Hooters—grounded in Americanized standards of feminine beauty—adapt the Hooters Girl in order to trade on female sex appeal in Latin America, and more specifically, in Colombia?

These adjustments/adaptations are key if Hooters wants to be successful beyond the cultural norms of the United States. As a profit-seeking business that is rapidly expanding internationally under the purview of globalization, the company must account for differences in linguistics, food preferences, and cultural understandings of female sexuality and sex appeal in order to be successful in Colombia—and everywhere.

Utilizing cultural artifacts related to the "Hooters of America" corporation, such as employee manuals, analysis of website and marketing tools, nonsystematic ethnographic observations, and public discussions and debates about the popularity (and suitability or respectability) of Hooters in Colombia and the United States, this chapter makes a comparison of how the corporation frames and enacts its restaurant concept in order to resonate with the cultural environment in which it exists. In particular, we address elements of female beauty, expectations of body shape, and differences and similarities in shaping a global brand that depends on women's bodies as a conduit for the organization's expansion. We use these to contextualize the similarities and differences of the two sites and to understand how it is that Hooters makes sense to members of both cultures, drawing them to the restaurants—and in some cases—leading them to protest the restaurants.

CONTEXT TO THE COMPARATIVE ANALYSIS: HOOTERS IN THE UNITED STATES AND COLOMBIA

Hooters opened their first restaurant in Florida in the 1980s, during the rise of the corporatized, sexualized fantasy woman in US popular culture. Most notably, the sexualized fantasy woman became the standard in the music industry (think Madonna) and was proliferated via MTV (think naked, writhing women in music videos). Female bodies became tightly controlled in the pop music industry, creating what O'Brien (2002, 238) calls the "fuckable fantasy woman," a "formula" that quickly spread to other popular cultural forms, and remains present to this day. Against this backdrop, two significant events happened that brought national media attention to the Hooters concept. In 1984, the day before the Super Bowl, John Riggins—then a famous player for the Washington Redskins—ate lunch at Hooters with several teammates, which received national media attention as a part of publicity surrounding the game. A few years later, the owners of Hooters submitted pictures of the first Hooters Girl, Lynne Austin, to *Playboy Magazine* where she became Playmate of the month and was also featured in popular Playboy videos. Thus, the Hooters Girl became embedded in the social fabric as a sexual commodity because it was linked to both sports and the zenith of the sexualized female body, *Playboy*. It is that sexualized commodity that the company continues to successfully trade on today. Indeed, Hooters' marketing materials state that Hooters—as a brand—is as powerful internationally as Coca-Cola, Google, and Apple, just to name a few.

The socio-political agenda of President Álvaro Uribe (who served from 2002 to 2010) aligned Colombia with the United States on economic and social (conservative) issues. This served as an incubator for cultural shifts that involved economic "prosperity" and opportunity. While its neighboring country Venezuela was shifting toward a more closed set of borders, with the imminent deployment of Hugo Chávez's democratic socialism, Colombia was opening up to the market and the possibilities of banking, tourism, and exportation of goods (coffee, flowers, and textiles, to name but a few). Once feared for its world-known coca production, the country's governance made a tactical shift, linking itself with a superpower in order to change its international reputation. As a result, corporations and capitalist elements (often submerged within democratic discourse) increased in Colombia.

At the moment when Hooters arrived in Colombia (first in Bogotá, and later in Medellín and Cali), it was a country less known to the world for being linked to the paramilitaries, and more as a place where "quality of life" and "peace building" includes having some wings while visually consuming female breasts. The Hooters-related social and cultural phenomenon in Colombia is less traceable to particular media moments than it is in the United States, yet, as we note, it is embedded in the opening of labor opportunities for women within these three cities—a relatively recent event in the contemporary history of this South American country.

GENDER, THE BODY, AND THE CIRCULATION OF DESIRE THROUGH BRANDING

We note that Colombia as a country is very diverse in terms of its population and its geopolitical location, and in its exposure to other political, cultural, and economic nation-states. The country is impacted by its Pacific and Atlantic coasts, as well as a periphery with five different countries. The native populations in Colombia range from people of Spaniard and Arab descent, to Afro-Colombians, people with indigenous background, and biracial or ethnically mixed Colombians. The regions in Colombia are coastal, mountain, and valley, ranging from sea level to 15,000 feet and higher. These different regions have various cultural, religious, and political positionings. Because of this, it is unlikely that one can say "this is how Colombians (as a whole) behave," since its regions are so different—from the coast to the Bogotá region to the borders with Ecuador or Brazil or Venezuela. Indeed, most of the research on Colombia is very localized and does not intend to represent Colombian "culture."

To complicate matters in terms of our comparative analysis, it is key to note that internal migration facilitates the interaction between millions of Colombians from all over—they often migrate to Bogotá, Medellín, and Cali, forming a complex notion of *colombianidad*,[1] given the mixture of peoples and traditions/ways of being and relating. Thus, our brief review takes into account work published in other regions of the country, since we view them as an important influence in the country's internal migration; we also consider economic and sociocultural realities from other cities to better situate the multi-city origins of people in the three cities where Hooters has been established.

Work on body, gender, and particularly women's bodies in international circuits of desire has illustrated the impact of homosociality (Bird 1996; Flood 2008) and men's use of language to link them through their desire for women (Kiesling

2005). Little has focused on the impact such homosociality has on the women, or on the men (based on their difference in socioeconomic status, racial identity, and relationship to hegemonic masculinity [Connell 2009]). We offer a succinct review of relevant sources in order to understand how cultural attributions of Colombian male masculinities and sociohistorical changes place Hooters in Colombia as a profitable and indeed sought-out space.

Research on male masculinities in various Colombian regions denotes a different set of explanations for the relationship between male masculinity and homosociality (see, for instance, Viveros 2002)—although, most often, these findings do not negate women's sexualized position in public spaces. In the coffee zone, in Manizales (Department of Caldas), scholars note the paradox of homosociality between heterosexual men, combined with enough distance, in order not to be perceived as homosexual, all while demanding a closeness to women, with the primary goal being the seduction and conquest of women (Palacio Valencia and Valencia Hoyos 2001, 67). Research shows how men's experiences with power shift as they reach adulthood: whereas as youngsters, they needed to compete against men in social situations such as sports, the shift as adults refocuses their attention to conquest. In Bogotá, research has looked at masculinity, gender, and sexuality, and their potential influence on intrafamilial violence, through comparative work between adolescents and adults in various neighborhoods within the capital (Jimeno et al. 2007, 94). The desire heterosexual men have for dominating women is compounded by the requirement that women must portray both innocence and availability. Women, on the other hand, are required to deflect their capacity to seduce, and assume themselves as the seduced. These findings were not representative of all neighborhoods researched, as in various sites, the men discussed the role women have as seductresses. More directly related to our chapter is the mention of specific meanings that people in a lower socioeconomic neighborhood gave to a billiard table, as a place where men's homosociality corresponded to the attribution of hypersexuality to women—to the point of seeing them as easy women, or prostitutes. The authors affirm that when women disrupt homosocial spaces, their sexuality turns them into sexual objects. Hooters might very well be another space where such attributions are made, though in this case, by people with higher socioeconomic status.

Arango Gaviria (2004) recounts the shifts in labor based on economic changes in Colombian society since the 1960s, and how these impact women's lives and families. Aside from emergent and increasing access to education for women, which their mothers in the 1930s and 1940s could not achieve, women after the 1960s were able to enter formal educational institutions. But there were economic shifts for women as well: having newer textile businesses offering jobs to women, which after the 1960s, afforded women not just temporary employment, but work they could sustain after marriage or having kids. What's key about Arango Gaviria's evaluation of these choices is the generational difference she notes: younger women's choices in the changing labor market implied individual choices; and for upper class women, work is seen as self-realization (just like for many US employees). Hooters women in Colombia might be accessing employment in that company for the possibilities of enacting a comfortable life in the city, or for paying for their studies (most universities require some form of registration payment—that is, education is not free).

Today, women comprise 70 percent of the workforce performing table service in the United States (BLS 2012). In Colombia, women comprise at least a third of the commerce, hotel, and individual and communal service workforce (Jaramillo and Castro Romero 2009). Prior to the success of the hotel and travel industries, it was men who were preferred as servers. Entrepreneurs quickly realized the benefits of hiring females as food service workers: cheap labor, reliability, obedience, tact, and attractiveness (Cobble 1991). Today, waitressing is both gendered and sexualized. Restaurateurs hire women for their nurturing tendencies (Cobble 1991) and often dress them in uniforms that highlight their bodies (Erickson 2004). Waitressing is a tough job: it is physically demanding with wage insecurity, often without other benefits such as health insurance or vacation/sick time. While many wait tables full-time, just like in Colombia, table service is an attractive job for young women, who use waitressing to sustain themselves while pursuing an education or as a stepping stone to other jobs. For them, the benefits of the job—money in their pocket and flexible schedule—are key.

We frame the sexual and cultural shifts through the economic change just outlined in order to better situate how Hooters, as a relatively inexpensive place to eat in the United States and an expensive place to eat in Colombia, has become so popular. Colombia, ready for an engagement with a global market, tourism, and a new beginning in human rights, attracted many US companies, and Hooters was no exception. But there are economic and social aspects that differentiate the use of Hooters as a space and as a way to access the bodies of women.

THE GENDERED, HOOTERIZED BODY

Corporate promotional materials tout Hooters Girls as local celebrities: glamorous, intelligent, and attentive as well as sexy and sassy. According to the company's chief marketing officer, Hooters Girls are the brand's "most precious asset" and the company conceives of the Hooters Girl as embodying the All-American Girl Next Door and/or the All-American Cheerleader. This is selling sexual fantasy—the wholesome, popular "girl" that men want to date and sexually consume. Hooters Girls occupy an interesting space in that they are both a "sexualized waitress" (Tibbals 2007) and a "waitress-entertainer" (Hearn and Stoll 1975); it is their bodies that are used to draw in customers, predominately heterosexual males. As a chain, Hooters must replicate this embodiment in every unit it operates and in order to do this, the company standardizes the appearance of the women who work as Hooters Girls.

The employee handbook, used in both countries, dictates appearance standards for the "uniform." The physical uniform, which the organization describes as "athletic in nature," consists of a white Lycra tank top and orange shorts.[2] The company is strict with respect to what cannot show: no midriff, no portion of the bra, no buttocks—all the while saying that the uniform cannot be too baggy. Sheer suntan pantyhose are also a required part of the uniform as directed by health codes, and the rest of the uniform consists of white athletic slouch socks, white athletic sneakers, a pouch with a belt that goes around the waist (for carrying money and other necessities), and a name tag.

There is much more to the uniform than material clothing, and the organization requires significant "body work" (Gimlin 2007) as well. The company mandates that hair should be worn with no unusual styles or adornments (large clips, headbands, etc.). It is implied that hair should appear to fall freely on the shoulders in order to create the appearance of glamour/sexiness. Feminine sexiness is also associated with makeup and the organization requires women to "accentuate their features" and be "camera ready at all times." Additionally, the company regulates jewelry, body piercing, and tattoos, dictating the actual number of rings and necklaces along with banning the showing of body piercings or tattoos. Manicured fingernails represent another area of eroticism of the female body and the employee handbook states that fingernails are to be "well-maintained" with no "extreme nail colors" or "excessively long nails." The organization is clear that the women should know the difference between too little and too much in order to strike the balance between too sexy and too wholesome—what others have referred to as the "Madonna-whore" continuum (O'Brien 2002).

The uniform extends beyond "looking the part" as women must also "play the part" when interacting with customers or representing the company outside of the restaurant. One of the major edicts in the employee handbook is to smile, smile, and smile! As with other interactive customer service work, employees are expected to project a friendly and cheerful attitude even if it is not consistent with their own feelings at that time, a behavior that is referred to as emotional labor (Hochschild 1983). At Hooters, the smile conveys more than just friendliness; it also serves as a flirting mechanism, which conveys sexual desire and availability.

Hooters imposes strict guidelines that sexualize all parts of the body (and emotions) as part of "the uniform" in order to standardize what is sexually desirable for the brand globally; from the United States to Colombia. However, there are different embodiment expectations in the United States and Colombia, which we turn our attention to next.

More Than a Mouthful: Tops and Bottoms

Colombia has an international reputation for its women making bodily adjustments and surgical reconstructions (with breast augmentation being one of the top surgical interventions as reported by the ISAPS 2011). Indeed, in work by anthropologist Michael Taussig (2008), we see how the unique emphasis on corporeal manipulations and surgeries operates in the lives of Colombian people: both in terms of the narco-traffickers, some of whom have covered up (and camouflaged) themselves through facial reconstruction, thus avoiding the law, and the beauty pageant contestants, whose bodies have been reshaped to compete with the neighboring country, Venezuela, still one of the top contenders in pageants like Miss Universe.[3] In Taussig's observations (mostly personal and travel accounts), Colombian men expand their bodies, sometimes developing incredible bellies and deformed bodies, but they pay for their female partners to beautify themselves, which results in one potential interpretation of the article's title, "Beauty and the Beast."

The cultural expectations of young women in Medellín or Cali, the second and third largest cities in Colombia, is to go through a rite of passage that sometimes involves breast augmentation as a *quinceañera*[4] gift. It is so engrained in the everyday life of women's negotiating space, access, and social mobility that a soap opera *Sin Tetas No Hay Paraíso* (literally, Without Boobs There Is No Paradise) was highly acclaimed and shown in and outside of Colombia, as it was (re)produced both in US Hispanic, and Spanish television.[5]

Most recently, people from outside the country come to go under the knife in Colombia, as the country is gaining a reputation for performing these services at a fraction of what it would cost in the United States or Europe.[6] For instance, a doctor known by the second author plans to develop spa-like services, known in other places as medical tourism, where people wishing to get any kind of surgery (cosmetic and noncosmetic) seek to blend the treatment and recovery time with vacationing and resting in hotel-like medical locations.

Like their Colombian counterparts, women in the United States seek out breast augmentation more than any other cosmetic procedure. While there is no television show that is a direct counterpart to *Sin Tetas No Hay Paraíso* in the United States, there is a pervasive cultural script that often makes women feel as if their worth and sexuality is directly proportionate to the size of their breasts. As such, female breasts are disembodied from the woman herself and become an object of and for male desire. In the United States, women experience low self-esteem related to breast size as their bodies are held accountable to the "Barbie doll ideal," described by Hesse-Biber as consisting of "exaggerated breasts, impossibly long legs, nonexistent hips, and a waist tinier than a Victorian lady's" (1997, 28–29).

At Hooters in both countries, it is clear that the women working the tables are young. Their ages range from 18 to 22—or 25 at most. In our nonsystematic ethnographic observations in both the United States and Colombia, we do not see breast augmentation as a critical aspect of the women's employment at Hooters. While there are women with implants, there are also women with nonsurgically enhanced breasts. In both places, the waitresses' breasts are often relatively "small" (or normal/proportionate), and often aided in size by noticeable push-up bras. While the women working for Hooters in the United States express concern about thinness and desire for larger breasts (as found in a study by the first author) this body focus shifts in Colombia, in spite of the branding of the company and its focus on women's breasts. In such observations, we notice the emphasis on the buttocks. In addition to their breasts, women draw more attention to their butts in Colombia, as the second author has observed, repeatedly, the use of foam (fillers, which are often noticeable) to expand the butts of several of the servers. In contrast to their Colombian counterparts, in the United States, the women do not emphasize being "bootiful," or highlight the buttocks, other than the desire to make them smaller or firmer.

Hooters is a double entendre. Owls, the official logo of the company, are called hooters for the sound they make, but the word is also slang for female breasts. This emphasis clearly resonates within the United States. It does not have the same meaning in Colombian culture, where embodied expectations about female sexuality center on the buttocks. This is not to say that the buttocks are not sexualized in the United States and that breasts are not sexualized in Colombia—our observations are that in US Hooters the emphasis by the women is on the top and in Colombia, the emphasis is on the bottom.

ACCESS TO THE HOOTERIZED BODY AS CLASSED AND GENDERED

In either country, the primary way to come in contact with Hooters Girls is in the restaurants. According to the company's website, there are currently 430 restaurants in the United States and abroad, with three restaurants in Colombia. At any given time, the company estimates that there are 18,000 Hooters Girls currently employed worldwide with more than 300,000 "alumnae." In the United States, Hooters is physically accessible, particularly in the Midwestern and Northeastern parts of the country. In Colombia, this is not the case. The restaurants are in the three largest cities: Bogotá, Medellín, and Cali. As such, Hooters in Colombia is more of an urbanized space, accessible only to those who live or pass through the space (including tourists from the United States and other countries).

Even if something is geographically available, price remains an important factor for access to the Hooters Girl. In the United States, Hooters price points are comparable to popular (and not particularly pricey) chains like Chili's, Applebee's, and Buffalo Wild Wings. In fact, Hooters is a popular choice for those seeking full-service dining, ranking 45th out of 400 chains in terms of sales in the United States (Restaurants and Institutions 2009). Eating at a sit-down restaurant like Hooters indicates, although not exclusively, disposable income. This is quite different in the Colombian Hooters. Colombian classes are clearly delineated geopolitically as well as through the services received. Every bill (electricity, water, cable, etc.) and document related to the services a Colombian receives certifies the strata or class level of that person; the lower (poorest) classes are Strata 0 and 1, while the highest (richest) class levels are measured with a Strata 5 or Strata 6.[7] Neighborhoods are laid out accordingly, with zones that are exclusively of the highest strata (some are gated communities, and, like in other countries, suburban). Hooters

Bogotá is in the Zona Rosa, or Pink Zone, which is very much Strata 6. This neighborhood is not a tourist zone, but may cater to those visiting or doing business in the city. The restaurant is a place where a person might spend $25 to $40 USD per person, which is not usually affordable for anyone from Strata 3 or 4 and below. Cali and Medellín's Hooters are both situated in Strata 5–6 neighborhoods as well. While it is not impossible to buy a meal at Hooters if one is not from the middle and upper classes, it is a challenge to those in Stratas 3–4 (and prohibitive to Stratas 1–2). It seems that Hooters in Colombia is a site that registers as upper middle class and upper class, and paying for food at a Colombian Hooters implies paying for status recognition.

These are not merely economic choices; they also carry sociocultural connotations. In the United States, Hooters is described on many websites as "low brow," which is also indicated in one of the company slogans: "delightfully tacky, yet unrefined." According to corporate marketing surveys of current and potential customers, nearly three quarters report that Hooters in the United States is seen as a place to gather to watch sports, let loose, and unwind. The casual beach theme clearly resonates with clientele in the United States and photographs on Hooters websites show people dressed casually (except at lunch during the weekday, when there is a mix). This is not the case in Colombia, where visiting Hooters is seen as a way to signify class. Because it is usually upper strata, customers are generally dressed more formally.

And, while men do like to go to Hooters, market research conducted by the company reveals they like to go with other "guys" but did not see it as a place to go with family or "girls."[8] Indeed, the women surveyed agreed that while they like to go to Hooters to watch sports, they do not see it as a place to take their family or to go with other women. Hooters, then, is a place for "guys." Our ethnographic observations and content analysis of Hooters websites echo this. While women and children are observed at Hooters, and appear in marketing materials, they do not constitute the typical guest party. Hooters recognizes this by calling itself a "neighborhood restaurant" and not a family one.

There are significant differences based on male masculinities, class, and social location (for work on Colombia, see Viveros Vigoya 2002). We seem to know much about males' use of space and their class; we know less about the ways in which women use space as it relates to their social class. Significantly, while men are thought of in terms of hegemonic and other types of masculinities (Connell 2009), women, on the other hand, seem to be placed in a homogeneous category (where class and race differences in relation to power are seldom explored). This is compounded by the fact that women's choices in working in service industries (Rambo Ronai and Ellis 1989; Tibbals 2007), whether erotic or not, are often reduced to ignorance about hegemonic masculinity—or power more generally. By focusing on this, we fail to explore the class aspects of women's experiences with such masculinities. While the experiences of women at Hooters in the United States have been explored, including how they locate class-based understandings of their work (see Loe 1996; Newton-Francis 2008; and Rasmusson 2011), we are left without much knowledge about women's choices in this kind of industry we call "body work" (Gimlin 2007).

EMBODYING *COLOMBIANIDAD*, SELLING NATION AS BEAUTY

In the United States, Hooters Girls are marketed as embodying the stereotypical European body type—constructed as white femininity. They are marketed as the Barbie doll ideal discussed earlier: blonde, busty, and lean. This does not meet with the realities of the women who work there. As the first author found (Newton-Francis 2008), most of the women stated that they did not look like the women in Hooters ads, noting that they came in all shapes, "color," and sizes. This sentiment was echoed by Rasmusson (2011) in her ethnography of Hooters.

Yet, there is a particular "look" and body type that Colombian women typify. The women at all the Hooters sites are not tall, which also offers a physical representation of Colombian women. Their height might be anywhere from 5′3″ to 5′8″. And while curvaceous, they are not large-bodied women either. Generally, they are skinny but with curves, and oftentimes have voluptuous rears—as discussed before, they were either born with them, or as we observed, fill them up with foam and other artifacts to make their bodies appear to cater to the expectation that men (and women) may have of Hooters Girls' bodies. Their hips and buttocks might be a sign of what folks in the global North think of as "Latina bodies" (Casanova 2004; Negrón-Muntaner 2004; see also Hill Collins 2004).[9]

And there is a mode, a common "type" of Colombian Hooters girl, which, we argue, typifies local, national, and international representations of Colombian women. While you almost always see a blond, or a black Colombian Hooters girl, our field observations reveal that the majority of the women at all three Colombian sites have dark eyes, black hair, and light-to-tanned skin. Their hair is almost always straight, black, and long, and their eyes dark as well, offering a more realistic regional beauty presentation. Because the mix of indigenous heritage and Spaniard blood is a main Andean form of racialized embodiment, being tanned or light-skinned offers a more regional or national look to the employees. As a case in point, in retrieving pictures of the first training conducted in Hooters Bogotá, two women—one from Guatemala, and one from Venezuela—trained Colombian women for the job in 2008. These women are both representative of a voluptuous, yet still skinny body, with long straight black hair, dark eyes, full lips, and with few indigenous phenotypical features. And they are both examples of the representation Colombian women face in the so-called beauty pageant market: Colombian women are often compared to their multiple-title-holding Venezuelan neighboring competitors, yet their presence and representation in pageants is a mixture of phenotypes that spans from Afro-Colombian and mixed-race women, to lighter skin (unlike the Venezuelan contestants, who tend to be chosen for their whiteness).

Guzmán and Valdivia (2004, 206) note that women "function as a sign, a stand-in for objects and concepts ranging from nation to beauty to sexuality." We want to add that globalization evokes more complex notions of womanhood, and Latina womanhood in particular, that signify beauty and nation, "here" and "there." As Guzmán and Valdivia explain, through globalization and tropicalization (Aparicio and Chávez-Silverman 1997), ethnicities are homogenized and identified as Latinas/os through a variety of stereotypical markers, most notably brown skin. In US popular culture, Latina bodies have been racialized, exoticized, and sexualized with different attributes being adopted as pornographic accessories, including long hair, curvaceous hips, large breasts and buttocks (Guzmán and Valdivia 2004). This popularized version of the Latina body is one that is being exported worldwide via celebrities such as Jennifer López, although there is a complex history of racializing and sexualizing the exotic, Latina woman in media that spans close to a century: the racialization of Latinas has taken place through the bodies of Mexican and Puerto Rican women in US media, starting with Lupe Vélez and Dolores del Río (from Mexico), and continuing with actresses such as Rita Moreno (Negrón Muntaner 2000), Rosie Pérez (Valdivia 1998), and most recently Jennifer López, all with Puerto Rican roots (Negrón-Muntaner 2004).[10]

More recently, Sofia Vergara (who moved from Colombia to the United States) has become the new "it" in terms of *latinidad*[11] as a form of sensual femininity in the United States. News media images (and text) foreground her curves, and interviews often foreground her sexual voracity. Her hair is long (it used to be dyed blonde but is now dark); her body has noticeably large breasts, and a significant set of curves/rear. Ironically, Vergara has been criticized for perpetuating Latina stereotypes when it is precisely those stereotypical characteristics that give her cultural caché in Hollywood. She represents the power of transnational racialization in that Hollywood (among others) is exporting this iconic body type throughout the world, including Colombia, where it has both cultural and monetary worth in the commodity culture. She is the most recent example of the domestication of the hypersexual (or at least sensual) and not so bright Latina bombshell. Way before the recent TV series *Modern Family* (which began airing in 2009), her work had already been perceived as stereotypical, in movies such as *Big Trouble* (2002) and *Chasing Papi* (2003), where she portrayed a Latina maid and a cocktail waitress, respectively (for a discussion on the former, see Soto 2008; for the latter, refer to Basler 2012). According to Fojas (2008, 427), the Latina "in classical Hollywood is often a counterpoint to the Euro-American female," oftentimes portrayed as a prostitute, harlot, or a dark lady (see also Basler 2012).

While we do not claim that Vergara's Latinization in the US media market had a direct influence in the expansion of Hooters in Colombia, we do see her moving to the United States after a very prolific career in Colombia, and her success in US markets, as a way to create a transnational figure of Colombian female beauty.[12] Colombianidad in this sense is intertwined with transnational fluidity and success in both geopolitical spaces, as it is based in the pride of being from Colombia. And like beauty pageants, actors, and media programming, sports are another arena where colombianidad surfaced in ways distinctive from constructions of US national identity.

Hooters is marketed as a place where sports are celebrated and watched. Most Hooters spaces have large TV screens with live and other sports to facilitate the homosociality expected of men who appreciate sports. At the time of this writing, the 2012 Olympics were taking place in the United Kingdom. When one visited the Hooters Cali Facebook page, the website offered all related programming, and an up-to-date count of the medals won by the Colombian athletes. Significantly, these posts are absent from the Hooters US Facebook site. Perhaps as important to note is the Afro-Colombian appearance of most of the Colombian athletes whose news and pictures are posted on Cali's webpage—eliciting a sense of pride that is not only national, but related to people of African descent (in order to draw more customers into the restaurant). In this way, the webpage functions to separate Cali's Hooters (as Cali has a larger Afro-Colombian population than Medellín or Bogotá; see Departamento Administrativo Nacional de Estadística, 2007) and situate it within a regional specificity that falls outside the mainstream parameters of Hooters in Colombia. At the same time, colombianidad helps homogenize a very diverse regional and ethnoracial Colombia.

NO TAN FELIZ EN HOOTERS (NOT SO HAPPY AT HOOTERS)

One of the company's slogans is "Hooters makes you happy." While Hooters has enjoyed a tremendous amount of success globally, it has not been received without acts of resistance.

As we have discussed here, Hooters is not simply transposed onto the rest of the world without some adjustments, and the "insertion" of Hooters elsewhere does not mean there are no resistance acts—not just to the perceived objectification of women, but to a globalized market reduced, in this case, to Americanization.

In the United States, resistance, generally on the basis of sexual objectification, has meant that the restaurants are picketed and the company criticized in both academic and popular literature. For example, on its website, the National Organization of Women (NOW) links to an organization that has called for people to mobilize and picket places such as Hooters that exploit women's bodies. Further, there are numerous op-eds in newspapers and examples of city council meetings where citizens voice their protest when a Hooters comes to town. Local governments have denied Hooters licenses to do business, as in Arlington, Texas, where the mayor stated: "It's the uniforms and the merchandise they sell, it's all very sexually suggestive. It's inconsistent with the standard of decency in this community" (Yee 2006).

Hooters recognizes this controversy and has produced an information packet for business and community partners titled "There Goes the Neighborhood, Now Better than Ever," to address some of these concerns. The publication highlights the casual atmosphere, great food, and service, and spotlights the various charities that Hooters supports. In an insert titled "Top 5 Misconceptions," they address sexual objectification, refuting the "myth" that Hooters exploits women. Their argument is that the uniforms are no more exploitive than NFL cheerleaders and that they provide a place for women to build skills that they later use in other careers. In fact, the company has started the "Orange Pride" campaign—which highlights former Hooters Girls that have gone on to become doctors, lawyers, entrepreneurs, and professors.

Nearly twenty years ago (in 1991), the Equal Employment Opportunity Commission (EEOC) investigated Hooters for denying men employment as Hooters Girls. In response to this charge, Hooters invoked the Bona Fide Occupational Qualification (BFOQ) section of the Civil Rights Act, which allows employers consider certain characteristics when making hiring decisions that in other contexts would be considered discrimination. Additionally, Hooters embarked on a media campaign (most notably billboards and newspaper ads) featuring a hairy man with a long blonde wig, in a Hooters uniform, with the slogan: "Come on, Washington. Get a grip." They also orchestrated a march on Washington. After several years of litigation, the EEOC stated that it would not pursue further legal action, allowing Hooters to continue to only hire women as Hooters Girls (Hooters of America 2007).

Hooters has also been the subject of academic critique, mostly from a feminist standpoint, which criticizes the company on the basis of economic inequalities related to patriarchy and male power—and "working for men" (Loe 1996), with some claiming that the restaurants' existence is "cringeworthy" (Rasmusson 2011). We find interesting that the kind of resistance that has existed in the United States toward Hooters is simply nonexistent in Colombia. Perhaps this has to do with the fact that Colombia was receptive to US business, that the business offers employment opportunities to women, or that, for the average Colombian, whose material needs are a daily struggle, having a Hooters is irrelevant to their making ends meet. Another potential reason is that there are few stand-alone units in the country (after all, there are only three, unlike the United States).

The only resistance we can find in Internet searches related to Colombia's Hooters is a scathing critique, published in a magazine for men (like *Maxim* in the United States) from Florence Thomas, a French feminist theorist who has lived in Bogotá for many decades.[13] As is to be expected, from a traditional feminist perspective, the article begins with a critique of patriarchy rooted in US/American ways of reducing women to objects. Thomas recognizes that as a business, Hooters brands everything: the space, the chicken wings, and women's bodies. By throwing women into the mix of what the company offers, Thomas effectively forms a complex platform to critique it. The critique is complex, in that the analysis of sexism is intertwined with the imposition of American values (or in this case, lack thereof). What is particularly striking about the comment by Thomas is the assumption that foreigners, not Colombians, consume the Hooters branding, women included. Furthermore, while Americanization and sexism are the most visible elements of this particular lens, there is little economic discussion about the prohibitive cost of the food for locals, as we discussed earlier. The chain is rejected based on the fattening quality of the foods, not the impossibility of the masses to pay for it (a critique that would more eloquently address globalization inequities). Yet Thomas expands the critique to the structural violence women face: "In this country of sexual violence, of gropings in buses, of everyday raping of girls, teenagers and women, of sad and sordid brothels, of women as spoils or weapons of war and of daily life, Hooters may seem harmless but it is not: it's the middle and high class version of a sexist and violent country." In this regard, we find her critique to recognize patterns of sexism and gender discrimination that merit further attention.

In her ethnographic visit, Thomas talked to the men, women, and children in the place, "fishing" for the answer she always expected—which was finally given to her by two men who said: "The chicken wings are good, but the girls are better." Assuming the women are cattle, and bunnies (she calls them both things several times), they seem, in Thomas's eyes, victims: assumed to be paid less than twice the minimum salary, and assumed to be working, dancing like spinning tops (*trompos* in Spanish), to pay off night college courses, the women are in the end portrayed as having no other choice. Something particular in the Colombian language is the use of phrases with double meaning: as a pertinent case to Hooters, "*comer*" (to eat) translates into "having" someone—usually, in street code, to eat is to penetrate or possess someone. To go "eat" chicken wings has thus a double usage. Thomas hints at this double meaning throughout her article, making emphasis on the objectification coming from the North. Similarly, in the United States "Hooters" has two meanings—it is a nickname given to owls and is slang for breasts. The symbolic aspect here is pretty obvious in terms of male desire.

CONCLUSIONS

In this chapter, we explored how Hooters, a US-based company, operates in Colombia, using culture and work, gender and the body, globalization and racialization as simultaneous lenses. Transnational corporations like Hooters—that trade on US understandings of beauty and the body—adapt to and are readapted by the culture in which they seek to do business. These adaptations are often embedded in complex sociocultural arrangements. In particular, homosociality is a common link to both Hooters in Colombia and the United States: that is, the men go to connect with other men through the service and perceived erotic labor of the Hooters Girls, irrespective of other variables; further research should certainly explore the motivations of customers (male and female) in patronizing such an establishment.

We argue that Hooters restaurants in the United States and Colombia call upon deeply ingrained and racialized understandings of female beauty. Hooters is successful in the United States because it privileges Eurocentric-based standards of beauty, while Hooters in Colombia calls upon deeply ingrained (and more *mestizo*) standards of beauty, which emanate from a less binary racial structure. This kind of comparative research outside of the United States offers different ways to think about racialization that move US scholars beyond black/white racial binaries. A parallel process of what some scholars call the Latinization of the United States (Negrón-Muntaner 2004) has made Latinos, and more precisely Latina conceptions of beauty, more significant and less peripheral to US cultural representations. This bodes well for the economic prospects of Hooters in Colombia, as there is a transnational circulation of desire with the franchising of Hooters there (and in other Latin American countries). It can be said, then, that as the Latinization of the United States continues to unfold, Hooters may trade on standards of beauty that are not so Eurocentric. This might mean fewer blondes, less focus on breasts, and a broader representation of female bodies. It may also mean more women of color entering Hooters' workforce. However, based on field observations in both the United States and Colombia, Hooters Girls are predominately white(r).

Our comparisons reveal that, while objectifying women is not new to the United States or Colombia, each does it in their own way. Like Thomas, the French feminist scholar based in Bogotá, we see the productivity of linking US violence (sexist, misogynist patterns) as something that enters Colombian culture—through wings—as a tool to help contextualize Colombian everyday violence for the reader who does not know the country. A feminist intersectional link that weaves the violence to the land and to certain bodies is indeed a welcome criticism of the cultural significance that could be given to Hooters in Colombia. We clarify, however, that both countries have patriarchal influences; furthermore, not only are they patriarchal, they are also classed. In Hooters, the middle-upper classes are the ones that enjoy the particularities of the female Colombian Hooters Girl, in contrast to US representations of working-class sexism toward the Hooters bombshell.

While we do not claim that in an era of globalization we can signal distinctive cultural patterns of meaning-making for the reception and representation of Hooters in the United States and Colombia, we see that racial and classed readings of the Hooters girl are different, and that the body parts emphasized, and beauty representations, are indeed different. Further research will explore whether the "fuckable fantasy woman" and the "All American Cheerleader" operate in both sites in the same way, or what differences stem from such a comparison. Certainly, not all cultural representations are received equally by all audiences, even within the same nation-state boundaries, so further work could also look comparatively at Cali, Medellín, and Bogotá and contrast their stores, and the experiences by the female employees at each of the sites (as one could do by studying the Midwest, East, West, and Southern parts of the United States to explore regional differences).

The links between the beauty industry and the media coverage of women's desirable assets in the United States and Colombia, the classed and racial complexities/differences, the body types and body parts, and the resistance to franchises such as Hooters all illustrate the place industries have in circulating desire and privileged forms of beauty, and expand on the notion of erotic labor in the service industry. As we have shown, a rather complex negotiation in the face of external commercial workings makes local adaptations to the Hooters Girl a way to utilize female beauty standards from the region. In the end, Hooters continues to proliferate through the male gaze and expectations that women eroticize their food consumption in ways that fuse the desire for a meal, and women, into one.

Notes

1. *Colombianidad* refers to the ways of everyday living and a sense of community among Colombians. For a development and critique of a sense of colombianidad, see Castro Gómez and Restrepo (2008).
2. The orange and white uniform is the most iconic, however, there is a black uniform that is worn on certain days or for special events.
3. Perhaps in this context Taussig simplifies the relationship of the passion for beauty to the passion for drugs, which for him means the emphasis on mafia and narcos that own stables and cars, and have beautiful women next to them as signifiers of their power. However, he emphasizes the potential role narcos may have in the contracting of plastic surgeons for preparing beauty queens for pageants (to perfect their bodies).
4. *Quinceañera* mirrors the rite of passage known as a sweet sixteen in the United States, but in Colombia, like other Latin American countries, this happens at age fifteen.
5. Let us not forget that *Ugly Betty*, the famous US TV series, was an imported version of (and not merely inspired by) Colombia's series. Oftentimes, Americans think *Ugly Betty* was imported from Mexico, given the copy of the series by that country. *Ugly Betty* in Colombia played with stereotypes of the non-pretty woman in order to show a revelation, a development of beauty that would resonate with the ideal of a beauty pageant from Colombia.
6. Medical tourism is an economic boon for Latin America and the Caribbean (see Ramírez de Arellano 2011 focusing on Cuba, Puerto Rico, Jamaica, and Barbados) and South America (see Edmonds 2011 on Brazil). For an overview of medical tourism in Colombia, see the following link: http://infosurhoy.com/cocoon/saii/xhtml/en_GB/features/saii/features/main/2011/01/03/feature-01.
7. See Departamento Administrativo Nacional de Estadística (or DANE) about strata and services: http://www.dane.gov.co/#twoj_fragment1-4.
8. The language of "guys" and "girls" was included on a survey by a consultant group hired to understand market demographics about current Hooters customers. The reported data were recorded in field notes with the notation that men were not referred to as "boys" but rather "guys," while women were referred to as "girls."

9. Interestingly, in *Black Sexual Politics*, Hill Collins assumes Jennifer López's body to be Black, and in her representation subsumes the image of the Latina sensual body into an African-descent category. While it is not uncommon for US African American scholars to highlight an Afro-descent Latino heritage as a form of sustaining a US biracial (black-white) structure, such provocative discussion goes beyond the scope of this chapter.

10. Carmen Miranda was a Portuguese-born actress whose influence was significant in US media; however, Portuguese women are not necessarily included in the "Latina" category as exemplified in contemporary ethno-racial politics; for a similar argument about Penélope Cruz (Spaniard) and Cameron Diaz (third-generation Cuban from her father's side), see Guzmán and Valdivia (2004).

11. She has also maintained a pretty high "mainstream" profile, perhaps as a result of her Latina background; in 2011, she ranked #3 most desired woman and in 2012 ranked # 1 (out of 99 women of all ethno-racial backgrounds) in askmen.com, a website that focuses on (presumably heterosexual) men's lifestyle and reaches about 20 million people. And her currency also denotes a central location in today's popular culture and media: according to forbes.com, Vergara was the highest-paid TV show actress/actor in 2012 (http://www.forbes.com/sites/meghancasserly/2012/07/18/sofia-vergara-modern-family-best-paid-actresses-television/).

12. It remains unclear whether Vergara's portrayals of Latina women are stereotypical or a sign of strength: whereas Mexican respondents spoke to Vergara's portrayal in the US television show *Modern Family* as a valuable Latina representation (successfully demarcating its transnational and global coverage—see González Alafita, Dávalos, and Gutiérrez 2012), students in Medellín viewed Vergara as a casual woman refusing to fit within a stereotype (Ruiz Marín, López Aristizábal, and Escobar Correa 2011).

13. This is the link to the article: http://www.soho.com.co/zona-cronica/articulo/florence-thomas-hooters/26085.

5

MOST DAYS I'M BEAUTIFUL

A REFLECTION ON SKIN AND BODY HAIR IN CAMBODIA
(*Personal Reflection*)

Kaija Bergen

Most days I'm beautiful. Like today, as I sit in the market next to an elderly little grandmother who grins toothlessly at me as she strokes my arm.

"Look how beautiful her arm hair is!" she hollers to the woman selling iced coffee nearby. Ming, or "Aunt," as most sellers are called here, nods appreciatively before frowning and loudly answering, "But she's gotten *k'mau!*"

The word "k'mau" translates to "black," but we aren't talking about black and white as we would understand it in the United States. In the Khmer language, to be "k'mau" describes being "tan" as well as all darker shades of skin color (though outside of skin color its use is limited to describing things that are literally black). It's a word I hear at least two or three times every day as people describe each other freely here in Cambodia.

Last year, having my arm stroked by an elderly stranger would have caused me discomfort. More than that, any public discussion of my physical appearance would have made me deeply self-conscious. However, living in Cambodia, I've learned to accept this kind of running commentary. For the past year of my service as a Peace Corps volunteer, almost every aspect of my body has been discussed: the pimple noted by my host sister, my weight

("slim like a Cambodian!"), my skin color (at first, "so white,"
now "black"), and, yes, even my body hair.

Responding to the coffee vendor's observation, I explain, "I
want to be 'black.' In my country, some people pay a lot of
money to make their skin this color."

But the women in the market are having none of my cross-
cultural explanations. "In Cambodia, it's good to be white,"
Ming says, and the grandmother next to me leans in to ask
conspiratorially, "Do you use a cream?"

No need to specify "whitening."

"No," I insist, "I like my skin. I'm happy with this color."

When I first arrived in Cambodia, I spent two months living
with a host family in a small village in Takeo province. The family
washed their clothes by hand in basins of rainwater, made feed for
their two cows out of fruit from trees in their yard, but bought
Nivea whitening body wash. Skin-whitening soaps and creams
are ubiquitous in Cambodia—so much so that it's a challenge
to find lotions without a bleaching agent. While the majority of
multinational corporations such as McDonalds, Starbucks, Wal-
Mart, and Shell have yet to enter the country, skin-whitening
products from corporations like Nivea and Ponds have found
their niche, entering the homes of the elite and poor alike.

I can't trace this value back to its origins, but it seems likely
that if French colonialism didn't introduce this idealization
of white skin, it certainly reinforced it. Although no one has
expressed any particular love for their former colonizers, little things like compliments about my nose being pointy and a
love of pale skin suggest that the French legacy extends beyond
baguettes and iced coffees.

Despite this colonial influence, cultures closer to home seem
to be fueling more of the white-skin obsession. While most of
the Cambodians I know aren't following much American or
European television, countries such as Korea, Thailand, and
Malaysia have a strong presence. Unlike Cambodia, these countries are much more connected to the Western media powerhouse, and I've watched countless adaptations of Western music
videos, movies, and commercials by these countries. It's not
unusual to watch Southeast Asian versions of Justin Bieber's
"Baby" and advertisements for creams and lotions featuring

pictures of famous white actors. Not to mention the Korean and Thai actors and singers that dominate Khmer television, who tend to be extremely pale. In what seems to function as a trickle-down effect throughout Southeast Asia, the love affair with whitened skin flourishes, reaching even remote Khmer villages.

My Khmer friends and host family do not use skin whiteners because they wish they weren't Cambodian or because they want to look like Zooey Deschanel or other pale Hollywood actors. Pride in Cambodian nationality is strong, and most Cambodians are unable to recognize even the most iconic American stars. In many ways, the value of whiteness comes from how difficult it is to attain and maintain. Here, white shirts are the "best" shirts because it's a feat to keep anything that clean in a dusty country where everything is hand-washed. In the same way, Cambodia's predominantly rural population means that those who can afford to stay out of the rice fields and merciless sun have the luxury of white skin. Being white, then, is more about wealth and socioeconomic status than ethnicity.

As it does most elements of Cambodian life, the legacy of the Khmer Rouge further complicates the politics surrounding appearances. Led by the infamous Pol Pot 40 years ago, the Communist Party of Kampuchea drove people from the capital and other cities into the countryside. Starvation and outright murder killed people from all demographics in the genocide that followed. Within their restructuring of society, the Angkar (as the Khmer Rouge leaders were called) favored the "old people." Not literally old people, but country people—the poor, rural Cambodians who had never "abandoned" their life of agriculture for the draw of urbanization and capitalism. Many of the "old people" were ethnically "pure" Khmer, with dark skin, curly hair, and limited connections with the paler, city people who might have Chinese or Vietnamese ancestry. Skin color was by no means the impetus behind the genocide, nor did it determine who would survive or be killed during the regime. Nevertheless, it could be a subtle factor in percep-tions of who was a "real" Cambodian peasant, and who had been evacuated from a more urbanized lifestyle. However, what this means for present-day Cambodia lies too far below the sur-face of daily interaction for me to say with confidence. Does an element of the association between dark skin and the ideal communist peasant continue to haunt Cambodia today, making people reluctant to trumpet the values of dark skin?

However convincing these ideas might be, I have to stop and examine my own cultural history. Perhaps I overemphasize the impact of the Khmer Rouge on skin politics, expecting a connection because of the role race has played in American history. Sometimes I think that I'm the one who is bringing my values of skin color and appearance to the situation, misreading comments because of what I have been taught to believe. I grew up in a family with an enthusiastic commitment to 1990s multiculturalism, playing with multiethnic dolls and watching shows about Kwanzaa and El Día de los Muertos, all of which seems to stem from the prevailing idea in the United States that skin color is about ethnicity and culture. It may be part of the reason why it's much more publicly acceptable to bluntly comment on appearance in Cambodia than in the United States. Perhaps it's because of my country's own loaded cultural history that these comparisons between skin tones often sound harsh to me. The majority of the time, when people in Cambodia talk to me about skin color, they aren't referring to ethnicity or culture. Particularly since any lingering politics of the Khmer Rouge go unaddressed publicly.

I'm reminded of the acceptability surrounding skin-color conversations as my Khmer friend Rathana and I look at pictures on her husband's laptop; she stops and points to a man.

"The black one is my husband," she informs me. "He's the reason my daughter is black. Black father, black daughter *(Pa k'mau, goan k'mau)*." Her practical tone implies a simple description of fact, and although I internally react to it, she doesn't perceive the comment as loaded.

Rathana loves her husband and daughter dearly, doting upon her little girl at every opportunity. Yet Cambodian social protocol permits very different conversations about appearances, with fewer taboos on direct comments about individual people.

Indeed, skin color isn't the only physical trait up for candid discussion in Cambodia, as almost all aspects of someone's appearance can merit comment in the market, at school, and at home. Moments such as the earlier market discussion about my

arm hair provides an example of conversations about appearances that, to me, feel too frank. "Sreymom is the short, fat
one, and Dara is the skinny, black (*k'mau*) one," someone
might say without being considered impolite. One of my more
awkward experiences of this candor had nothing to do with my
pigmentation.

"Why is there no hair?" my host brother inquires out of
the blue one evening. I'm confused until he gestures towards
my legs. My host sister, along with her two teenage children
and grandmother, all turn from the TV to listen to my answer
intently. The mystery of my lack of leg hair trumps their Thai
soap opera.

"I, uh, shave it off," I explain to blank looks before expanding my answer, "like a man shaves his face."

This addition apparently does nothing to clarify the situation.
"Why?" my host brother asks, bewildered. Though bleaching
skin is common practice here, shaving for women is not. While
(or perhaps because) I don't support the former, I find myself
struggling to justify the latter. My answer feels inadequate from
the start. "Um, women often do this in the United States. We
think it looks better."

"It doesn't," my host brother informs me with assurance.
"It's better to have hair."

His response echoes the attitude of the women in the market
regarding my arm hair. I'm baffled. No one has ever complimented me on my body hair in the United States. The ghost-
white Korean pop stars that dominate my host family's television,
as well as any American film icons, may be baring shaved legs,
but the standard doesn't seem to have impacted the Cambodian
people I know. My host family isn't alone in their confusion
about my hair removal—other Cambodian friends have con-
fronted me about it as well. The unpopularity of shaving makes
me reconsider my theories on the desire for white skin. Do I
assume people want white skin because I have white skin? Why
would that standard of Western beauty gain traction when others have so clearly not? In a country where most people are relatively body-hair-free in comparison to someone of Scandinavian
heritage like me, body hair is desirable. In a country where sun
exposure darkens skin, "white" skin becomes the goal.

What translates from one culture to another? What perspectives come through the television, the movies, the product distribution, and what do not? What points of view were already there? Sitting in the market or at home with my Cambodian family, I'm most aware of the degree to which I don't understand where values of beauty originate. Maybe these standards stem from French colonial values, vestiges of a one-hundred-year legacy that continues to impact people's perceptions almost fifty years after the *barangs* (the French/Westerners) left. Or perhaps they are a reaction against the history of the Khmer Rouge, a disavowal of the demonization of "new people." Maybe they come from China, Korea, or Vietnam, countries whose immigrants, television shows, and products are all entering Cambodia at an accelerated rate. The lines between "authentic" Cambodian values, colonial values, Western globalization, and Eastern globalization are all blurry. This isn't even a complete list. As the ongoing discussion of my body hair indicates, "beauty" is far more complicated than any one influence. People don't just want to be what they see on television or believe to be "American" or "Chinese" or the stereotype of some other cultural powerhouse.

I'm tempted to draw conclusions, and I think that the way I have been taught to think about skin color lends itself to this instinct. After all, I know white people come from the West, the West has a powerful history of forcing its ideas and values on other countries, and therefore it would be easy for me to believe that Cambodian skin politics stem from the West. However, in an era where technology allows complicated exchanges across the world, I'm not sure that these values are as easy to trace. Now, more than ever, people in Cambodia and around the world are able to observe, appropriate, and engage many other cultures. I don't necessarily know what it means to be "white"—in the United States or in Cambodia. The more I attempt to determine the impetus behind preferences in appearance, the more arbitrary all of it seems. Some days in Cambodia, I try to understand what makes me beautiful, but I think the truth is I'll never fully know.

6

REPRODUCING BEAUTY

CREATING SOMALI WOMEN IN A GLOBAL DIASPORA*

Lucy Lowe

The first day I met Sumaya,[1] a young Somali woman who would later become one of my closest friends in Eastleigh, she slopped a face mask on my skin, pondering what would happen if she used the skin-lightening one on my pasty Scottish face, ran oil through my hair, and gave me a brightly colored *dirac*, a very popular style of Somali dress. These dresses, worn by women of all ages, were fairly universal in size and somewhat shapeless, but worn with an underskirt, the sides could be tucked in to create a flattering, draped waistline. At the time I found it a little odd that this woman, just a few years younger than me, had decided that it was necessary to put me through a rapid beauty regimen on our first meeting, but I enjoyed the pleasant smelling lotions in contrast to the grimy, dusty streets outside, and as I was to discover, this sort of thing was to become a regular aspect of my fieldwork.

LITTLE MOGADISHU

More than 20 years of conflict and occasional droughts, floods, and famines have sent over a million Somalis into neighboring states and beyond, and by the end of 2011, Kenya was host to over half a million Somali refugees (Lindley 2011). During my fieldwork, Somalis were given *prima facie* refugee status, granted on the basis of their nationality, rather than through the determination process that asylum applicants from other states were required to follow. Due to the encampment policy,[2] all refugees in Kenya were required to stay in one of the refugee camps, located in remote, arid parts of the country. Despite this, many people chose to leave the camps, or bypassed them altogether when arriving in Kenya, going directly to Nairobi or other parts of the country, with tens of thousands settling in Eastleigh, a neighborhood situated a short distance to the east of central Nairobi.[3]

Eastleigh, commonly known as *Mogadishu Kidogo* [Little Mogadishu], is notorious throughout the Kenyan capital and beyond. The residents are predominantly migrants, mainly from Somalia, although there are also many from Ethiopia and some from Eritrea and Djibouti. As a result of this, the neighborhood is rumored to be a hive of pirates and Islamic fundamentalists, a refuge for people illegally present in the country. Interestingly, Eastleigh also has the reputation of being one of the best shopping areas in Nairobi. The emergence of the area as a Somali-dominated neighborhood has transformed it into a commercial hub, with enormous, shiny shopping malls rapidly emerging along the dusty, potholed roads. Within these malls unfold immense rabbit warrens of small shops, selling everything from clothes to electronics. The quality of goods and the cheap prices draw consumers from throughout Nairobi, and even some wholesalers from neighboring Uganda and Tanzania, despite the rumors of pirates and terrorists.

The economic prosperity of Eastleigh's business community has contributed to the perception of Somalis as wealthy, and their legally dubious presence has made them an easy and alluring target for police seeking to supplement their wages with bribes (Human Rights Watch 2010). The area is notorious for widespread insecurity, with violent and often armed robberies of individuals and businesses being a common occurrence. As a result of this often hostile context, and with limited opportunities for the future, the overwhelming majority of my informants aspired to move abroad, particularly to North America or Europe. Amina, a woman in her mid-twenties, who had spent eight years in Kenya, explained to me:

The camps are like prisons. There is nothing to do, there are no opportunities. They keep refugees in them like criminals, but we did nothing wrong. They don't let us travel anywhere or do anything unless we pay them [in bribes]. So that's why everyone wants to leave this place. It's better for Somalis to go outside [abroad], there are opportunities there, you can get a good education, work. And without the insecurity we have here.

Twenty years of this desire to migrate "outside," either through claiming asylum, family reunification, resettlement by the United Nations High Commissioner for Refugees (UNHCR), or using one of the various illegal methods of crossing borders, has created a vast diaspora of transnational family networks. I use the term "diaspora" not to suggest that they are an entirely homogenous community (Ong 2008), but to describe the ways in which many Somalis, forced to flee their country, have tried to maintain both connections with their extended family and in many cases, a spatial unity in their settlement patterns, such as the Somali enclaves in Nairobi, London, Minnesota, Helsinki, Toronto, and Johannesburg, among many others. Ong argues that "there is a polarization between those free to move and those forced to move" (2008, 171), and while this is true in many cases, the ongoing, multidirectional migration of many Somalis blurs this distinction between what is "forced" and what is "free."

The Somali diaspora thus provides a fascinating way of looking at and thinking about globalization, particularly if we step away from the neoliberalism that is so frequently synonymous with the term (and which has undoubtedly affected the current situation in Somalia and the experiences of refugees), and instead focus more on the growth and intensification of the rapid transnational spread of communication, goods, and, of course, people (Edelman and Haugerud 2004; Friedman 2004; Inda and Rosaldo 2008). The length of their exile has resulted in a new generation of adults who have been brought up in foreign countries, many of whom retain the desire to return "home," even if they were born elsewhere. If we think about globalization as the rise of global movement and interactions, and the decline of the power of the nation-state, Somalis, having not so much a decline as a complete rupture of the nation-state since 1991, provide a unique example. The power of the United Nations and international nongovernmental organizations (NGOs), both within Somalia and over the lives of many Somalis living abroad as refugees, is illustrative of how far removed they are from the influence of their own nation-state. This distance is even more evident among those people who have fled illegally over international borders, and for those who regularly send and receive remittances to and from family all over the world through *xawaala*.[4]

I carried out 20 months of fieldwork in Eastleigh from November 2009 to June 2011, primarily looking at how practices and beliefs relating to childbearing had been affected by displacement and subsequent migration to other countries. During this time I spent six months volunteering in a busy maternity hospital, three months volunteering with a Kenyan NGO that provides legal advice and support to refugees, and spent the remainder of my time carrying out formal semi-structured interviews (with 45 women and 17 men) and less formal, unstructured interviews (with more than 120 women and more than 40 men) as well as participant observation within the community of Eastleigh and in relevant NGOs. The majority of my informants were between the ages of 18 and 40, although an indifferent attitude toward birth dates and precise numerical ages meant that informants often gave a vague idea of when they were born or what age they were, while others changed their age on the different occasions that I met them. Most of my female informants were married for the first time by their late teens or early twenties, while my male informants were normally a few years older when they married. High divorce rates among the Somali community in Eastleigh meant that it was fairly common for informants in their later twenties and older to be divorced and in many cases remarried. The scale of displacement and onward migration, not to mention the restrictions imposed by international immigration regulations, meant that married couples were frequently geographically separated from each other, with one spouse in Nairobi and the other in Somalia, a refugee camp, or a third country.

Although the main focus of my research was reproduction and maternity among Somali women, as my fieldwork went on

I found myself spending increasing time in salons or in women's homes, discussing and practicing "beauty." This may sound like a leap in topics; however, this was not the case. Homes, beauty salons, and hospital waiting rooms were the three main social spaces for women during my fieldwork. Within them, women were able to discuss their lives while comparing current trends in fashion and beauty (Ossman 2002). In doing so, these women were renegotiating ideals of "being Somali" while displaced as refugees within Kenya. By sharing and maintaining practices that they associated with memories and ideas of their home country, they were engaging in what Edmonds termed "aesthetic nationalism" (2010, 41). Tiilikainen suggests that perpetuating what are perceived as traditional or cultural duties helps Somali women in Finland to maintain "a sense of continuity, order and control in the middle of otherwise chaotic experiences and uncertainty" (2007, 212). Furthermore, Akou argues that "for Somali refugees, a strong sense of collective identity—projected through clothing—is almost all they have left of their nation. Unable to return to, or in many cases even visit their homeland (which is still involved in a violent civil war), Somalis use clothing to keep their memories and dreams alive and to shape the future of a new Somali nation" (2004, 51).

As this chapter will examine, in Eastleigh it was not only the use of Somali clothing, but also active attempts to produce and maintain what were perceived as beautiful Somali *bodies* that were an evident aspect of reimagining a sense of Somali identity in the diaspora. At the same time, the widely geographically dispersed diasporic community meant that these Somali women were attempting to maintain such ideals while in a context of rapidly globalizing encounters. As such, ideas of "Somaliness" or what a "Somali woman" is or should be, or how best to be one, were being reshaped within the context of displacement and perceptions of their host country and community (Cvajner 2011; Isotalo 2007; Tiilikainen 2007). For Somali women in Eastleigh, maintaining ideals of Somali femininity and beauty were evident as attempts to retain a sense of cultural identity. In this chapter, I will explore how the ideals and concerns surrounding beauty and what it means to be a "good" Somali woman shaped the daily lives of my informants in Eastleigh. In doing so, I will also examine how these practices and perceptions were influenced by their diverse experiences and interactions with an increasingly transnational community, thereby renegotiating what it means to be a Somali woman in a globalized world.

Social Beauty

Although they were certainly not what most people associated with Eastleigh, beauty salons could be found on almost every street. Some streets, in fact, had several of them, nestled together in a line, their entrances and windows covered by curtains to hide the goings-on of the patrons inside. The interiors were decorated with large mirrors and images of women, including photographs of traditional and modern Somali bridal attire, Bollywood actresses, and African American singers. Skin and hair creams, oils, and dyes were stacked up on shelves, as well as smaller trinkets such as glitter and hair clasps. Although *uunsi*, a Somali incense, was often burned in salons, its heady smoke was powerless against the overbearing smells of hair peroxide and the large tubs of henna.

Salons, owned and operated by Somalis, Kenyans, and Ethiopians (although my Somali informants visited Somali-run salons almost exclusively), provided both a source of income for women and a space in which they could socialize and exchange thoughts and advice on aspects of their lives ranging from hairstyles to infant feeding. As salons were exclusively female spaces, with the occasional exception of an awkward-looking young boy forced to accompany his mother or sister, they were in many senses more private than homes, where men could enter and exit whenever they pleased. Men would sometimes escort their wives or sisters to the salon, but were only expected to make the briefest of greetings before disappearing. As they were such small rooms, it was usually only possible to fit a maximum of four or five women, plus one or two staff members, in them at a time. This created an enclosed and intimate space, away from prying eyes and ears. Most of the beauty treatments performed in salons, whether applying henna, face creams, or hair relaxers, took considerable lengths of time, and it was therefore very common for women to spend several hours on each visit.

Face creams, moisturizers, perfume, skin-lightening creams, jewelry, incense, hair treatments, hair removal products, and makeup in bright, colorful packaging were just some of the ubiquitous products aimed at maintaining both aesthetic and olfactory beauty that were stacked up in shops and stalls lining Eastleigh's murky streets. Beauty products were easy to come by, and although the wide assortment of products varied in price, most were fairly inexpensive, making them accessible to most women, and certainly cheaper than frequenting salons on a regular basis. Many young women I met who had little or no income would often use their sisters' or friends' products, and as a result of this socializing of beauty rituals, there was very little difference between how they were enacted in the home and in the salon.

The routine of applying creams, oils, and lotions to the face, hair, arms and hands, feet and legs, was something I witnessed almost every day. After breakfast had been prepared and consumed, the women I lived with began the process of cleansing, moisturizing, and beautifying, a process that took most of the day. Hair removal was seen as a further aesthetic requirement, but one grounded in religion,[5] and most young women regularly removed body hair with epilating creams such as *Veet* when necessary. Although some older women said that in Somalia they had used a thick, sticky solution to pull out the hair, young women told me that was "old-fashioned" and that *Veet* was "easier" and "more modern."

The women I lived with were certainly not unique in their observance of body care, and the five-story apartment building I lived in was abuzz during the day with women moving between each other's homes, making tea and cooking in the open hallways, all the while with masks of green, white, and orange coating their faces, and thick socks holding on the plastic bags which protected the moisturizing creams covering their legs and arms. Even in public in residential areas I often saw a woman popping out to a nearby kiosk with a mask on her face. I quickly noticed that if I looked closely at women wearing *niqab*[6] on the street, it was not unusual to see a slick of green or gray mask around the edges of the eyes. I mentioned this to the women I lived with, who said they sometimes did the same thing, and the youngest of the sisters, who suffered from acne, also said she wore the *niqab* when her skin was particularly spotty.

Women explained to me that maintaining soft, beautiful skin and hair were essential to attract a husband, and subsequently demonstrate love and affection for him. Although parents, and particularly fathers, continued to play an important role in arranging marriages, the dislocation of families through displacement had resulted in an increasing number of young people meeting and selecting their own spouses. While both men and women informed me that they were able to refuse the partners that their fathers suggested for them, in practice it was largely the men I spoke to who had felt able to decline. As such, men were often in a far stronger position when it came to decisions regarding marriage, and the "beauty" of potential wives was often an important factor. In her discussion on Palestinian women, Kanaaneh (2002) describes how concern for taking care of the body has shifted toward a deep concern with physical aesthetic appearance. Similar to what I encountered in Eastleigh, she found women to be deeply troubled by how pregnancy and breastfeeding affected their bodies, placing emphasis on their sexual appeal (2002, 175). Kanaaneh suggests that rather than this being a move away from family ideals, it is in fact incorporated within the pro-family context. In Palestine and Eastleigh, women were eager to be attractive to their husbands, and thus ensure the future stability of their marriage. In both cases, imported products that were thought to help maintain an attractive physique were often perceived as more effective than "local" or "traditional" alternatives.

Saido, a woman eight months into her pregnancy with her third child in as many years, invited me to her home to see her latest purchase—a large elasticated band that fits snugly around the waist. As she inspected the box, featuring a smiling blonde woman wearing the band, she explained to me that she intended to wear it after giving birth to her child so she would look "slim and beautiful, like I did before I had any children. Like Angelina Jolie!" I asked her what she had used after her first two pregnancies, and with a dismissive shrug she answered, "I tied some fabric around my belly. This will work much better." "Why?" I asked, "Everyone does the same thing here, Somalis, Kenyans, and sometimes it works and sometimes it doesn't. It might work to begin with, but the more babies you have the more difficult it gets. This is from America, so it must be better."

Although Saido, like many young women I spoke to, was maintaining what I was told was an old practice of binding the waist in an attempt to regain the figure she had before she became pregnant, she was utilizing imported goods that she perceived to be more effective. A more significant deviation from 'traditional' practices was the growing popularity of diet pills. Frequently advertised on the Egyptian television channels that were available in the area, the pills were also evident in the posters that plastered the walls of the many pharmacies and clinics in Eastleigh. Although only a very small number of my informants told me that they used them, those who did said they had never tried anything like them before they had come to Kenya, and two women told me that they had been recommended to them by their sister living in America. Both the elasticated band and the pills were incorporated into preexisting ideals of how young Somali women with few or no children should look, with a primary emphasis on appearing beautiful for their husbands.

Interestingly, although a slim physique was considered attractive among young women, exceptions to this rule were made for women who had given birth to a significant number of children. It was generally deemed unreasonable to expect women who have provided their husbands with children to maintain a slim figure. This situating of beauty as something relative to the particular stage in the life course is not by any means unique to Somalis, but what was important was how beauty, and whom one should be beautiful for, is dependent on the individual's social position within the family. Wives should be beautiful for their husbands, but mothers often expressed a greater bond of love between themselves and their children, rather than their husband. As such, women were often complimented on their beauty by their children and even grandchildren, altering one's role from being a beautiful wife to being a beautiful mother. With the exception of ideal body shapes for women depending on whether or not they had produced children, the standards of beauty did not change hugely—soft, perfumed skin and hair— but the intended recipients did.

MIGRATING PERCEPTIONS OF BEAUTY—MAKING
THE GLOBAL LOCAL

The rise of communications has had a massive impact on the lives of people in Eastleigh, especially if we consider, as someone explained to me, that it was not so long ago that people were using two-way radios to communicate with relatives in Somalia. During my fieldwork, most adults had mobile phones with Internet capabilities, and access to email, social networking, and instant messaging, hugely simplifying communication with family, wherever they might be.

Nairobi is an internationally well-connected city, one of the largest, safest transit points en route to and from Somalia, and host to one of the largest Somali settlements outside the refugee camps. Although some residents had been there for 20 years, others used it as a stepping-stone on their way out of Somalia, while yet others had moved back and forth between Kenya, Somalia, and other countries inside and outside Africa. As such, concepts including beauty and tradition were constantly being reassessed and reimagined, as the relocation of bodies carried with them ideas, images, and products (Hernlund and Shell-Duncan 2007, 3).

Due to the insecurity, as well as the perception that "good" women should stay at home and not be seen hanging out in the streets, my informants spent a significant amount of time watching television, which in addition to access to the Internet, thrust them into a readily available global world of new goods and images. The cable in most homes was illegally hooked up to one of a number of local providers, who were able to determine which channels were available to their customers, resulting in viewing options being strictly controlled and limited to "appropriate" channels, particularly during Ramadan. At one point during my research, a Spanish-language soap opera was available; however, the very mild sexual content caused outrage among local religious leaders, and the channel swiftly became inaccessible. The American channel *E!* met a similar fate. Fatuma, in her early twenties, complained to me,

> These *sheikhs*[7] have nothing better to do but complain about some kissing on TV. There are plenty of things they could be

doing—what about the thugs running around here, these young boys who are fighting and stealing and causing trouble for the rest of us? No, all they want to do is complain about what we watch on TV.

For their part, the sheikhs often claimed to be "protecting" the community from detrimental external images, as one told me:

It's not good for our young people to watch these things. If we were still in Somalia they would never see such things, and they would have their extended families to keep an eye on them and make sure they are behaving. But here, because of the problems in Somalia, they don't all have that, and they are exposed to all sorts. It is our duty to make sure that our people don't end up becoming immoral and running around the streets.

The television options that were available included several Somali language channels, some broadcast from Somalia, others from the United Kingdom, a number of Arabic channels, often showing American or British movies, made suitable for viewing by having their scenes of sex and nudity edited out, and the immensely popular Indian channel, Zee TV. The Somali channels were particularly popular, showing Somali music videos, some of which were filmed in Somalia, but most were made within the diaspora, in Kenya, Europe, and North America. They featured singers sporting the latest fashions, often influenced by where they were shot, but still with a distinctly Somali style. These channels also featured programs specifically aimed at women, such as those providing religious advice on family matters and those promoting beauty, such as *Qaab iyo Qurux* [Shape/Figure and Beauty], a show, unsurprisingly, about fashion and physical attractiveness, where viewers from throughout the diaspora were able to phone in and ask questions. Although filmed in London, the program is available to Somalis internationally through cable television and the Internet, and during the show it was not unusual to have one caller from Manchester in England, and the next from Hargeisa in Somaliland. This transnationalism brought women living in diverse parts of the world together, to exchange and engage in ideas of what and how a Somali woman should be, and significantly, how she should attain these ideals.

The Zee TV channel and Bollywood DVDs were also very popular, to the extent that many of my informants (mainly women) were able to understand Hindi. The influence of Zee TV stretched well beyond simple entertainment, as women would admire and often seek to reproduce the styles of clothing, hair, makeup, and jewelry that they observed on television. For most of my informants, daily clothes were simple; their hair was tied up and covered with a scarf for most of the day, and makeup was minimal, if used at all. Weddings, which were often extravagant events, gave women the opportunity to experiment with their appearances, and display them in a context and manner that was safe and respectable. As weddings were always evening events, women had the entire day to spend on perfecting aesthetics. At almost every wedding I attended, there was at least one woman wearing a *sari*,[8]and it often caused ripples of conversation as to how beautiful they were, but at the same time, whether it was appropriate for a Somali woman to wear one, "She should be wearing traditional Somali dress, not Indian, who does she think she is?"; "I heard it's *haram* [forbidden] to wear them"; "She looks nice, but our [Somali] dresses are so much more beautiful." Such comments and discussions were evidence of the tension women felt between experimenting with the foreign goods that they saw and often admired on television, and the feeling that as Somali women, they should be maintaining links to their "own" country and culture. Interestingly, I never heard similar comments made about young men, who largely rejected "traditional" attire, preferring to wear the fake designer Ralph Lauren and Gucci clothing that was imported from China and widely available in Eastleigh.

It was not only the guests' attire that was the subject of such discussions, as the brides at the majority of the weddings I attended wore what were described as "Western" white wedding gowns. A fellow guest at one wedding said to me,

It's the fashion now. Some women wear the traditional styles, but people like this. It's what they see in the movies, and they want to have weddings like those ones. I don't think people even think of this as being foreign anymore, especially when the brides are like this one, covering their hair and making sure they still look modest. *Masha'allah* [praise to God], she looks beautiful.

So maybe the dress is Western, but she wears it like a Muslim, and she has henna [on her hands], so she looks Somali.

In contrast to most women's everyday appearance, bridal makeup was normally applied rather heavily, beginning with a thick coat of pale foundation, making the bride appear to have a much lighter skin tone than she actually did. Added on top of this were lipstick, blusher, eye shadow, mascara, and eyeliner. Henna, carefully painted on the hands, feet and sometimes the chest in intricate patterns, was also essential to a bride's beautification. Women would use henna for any occasion, when they could afford it, but the most elaborate designs were always reserved for weddings. It could be purchased quite cheaply in the markets and applied at home, and it was the earliest form of beautification I observed girls practice, as they would paint it on their hands and feet, or have their mother or sisters do it for them. Women sought more ornate designs in salons, where intricate patterns were often drawn all the way up arms and legs. Such bodily décor was greatly admired by men and women alike, with the latter particularly aware of the latest styles, often inspired and imported from abroad.

Young women were readily able to distinguish between what was a new, fashionable design and what was deemed to be dated, with the difference often barely detectable to the uninformed observer. When I asked where they got the latest fashions from, many said they had seen them on TV, but often it was from a relative or friend who lived or traveled abroad and had imported them. One young woman in a salon told me,

I see photos and sometimes wedding videos of my family in Dubai and in America, and you know how Somalis like to look nice for the camera [laughing]! So you can see their style and how they do things, and if I like the way they're dressed or how they have their hair or henna then maybe I'll try to do the same thing. And when they come to visit they bring things for me, like new clothes or makeup, and then I can enjoy the nice things they have out there. That's how we learn about the modern fashions.

Wedding videos were immensely popular, as they were perceived as necessary in order to prove a marriage was real for immigration purposes, and many people believed that they were more effective than marriage certificates. At one wedding I attended, the groom was present but the bride was in Minnesota, and represented at the ceremony by her uncle. When I later watched the video, the celebrations that had occurred in Kenya and the United States had been mixed together, depicting a truly transnational wedding. These videos were frequently shared among women, and featured Somali weddings taking place in a diverse range of settings. While watching, women would admire (or criticize) what they saw and heard, and like the other images they saw on the Internet and television, they often tried to replicate the ones they deemed to be desirable.

Perhaps one of the greatest exposures to globally mobile concepts and goods was in the movement of people. A number of my informants had lived in other countries, particularly the United Arab Emirates (UAE), the United States, the United Kingdom, and Yemen, and a smaller number in Egypt, Saudi Arabia, Norway, and Sweden, and everyone I spoke to had at least one relative living "outside," although they were not always in contact with them. The mobility and highly transitory nature of the community in Eastleigh meant that the movement of bodies and goods was both frequent and rapid. Some families set up transnational businesses, and imported goods into or out of Kenya, but most maintained contact through sending remittances, and thereby providing financial support. When relatives came to visit they were expected to bring gifts, and during my trips back to Scotland I was given shopping lists including perfume, makeup, dresses, jewelry, and skin creams. The foreign products imported by myself or others, whether they were electronics, clothes, medicines, or beauty products, were perceived as more effective than anything that could be bought locally, even if it was the same brand. "Can you bring me some Nivea cream when you come back?" one friend asked me while she watched me pack my suitcase. "But you can buy Nivea here," I responded. "But yours is better," she informed me matter-of-factly. Products brought from abroad had the added prestige of being something that other friends were unlikely to have, and yet the beauty items they desired and requested were generally used to conform to "Somali" concepts of beauty, clearly blurring the lines between "local" and "global," and "traditional" and "modern."

BECOMING A WOMAN, WIFE, AND MOTHER

As mentioned earlier, when considering what beauty is, it is essential to think about whom or what beauty is for. Among Somali women in Nairobi, it was impossible to disentangle the meanings attached to the aesthetics of the physical body from the roles which were considered essential aspects of becoming a woman—marriage and motherhood. The importance of child-bearing and high fertility was stressed to me throughout my fieldwork, not least by my friend Muna, who took great delight in reminding me that although we were the same age (27 years old), she had ten children and was pregnant with her eleventh, while I was yet to have one:

> I don't know why you white women wait until you are so old to have children. It's not good for you, and if you wait too long, you might not be able to have them at all. After thirty it becomes very hard, especially if you haven't even had one by then.

Muna's reproductive achievements were not only a reflection of her success and status as a mother; they were a very visible illustration that she was a "good wife" (as her husband proudly told me every time the topic of childbearing came up) and a "respectable woman," capable of producing and caring for a family. Through marriage she had established her own family, and by producing children, particularly such a large number, she had secured her position within her marriage with her husband as well as his extended family, all of whom she could reasonably rely on for support in the future, should she need it. In the hostile, transitory context of Eastleigh, the women I spoke to were highly aware of how important marriage could be to their own physical and economic security, as well as their possibilities for onward migration. The risk that one's husband might leave and find another wife was felt even more acutely within the number of transnational families, where spouses were often living thousands of miles apart. As one woman told me during a conversation about her husband's imminent departure for America, "I only have two children. I think he'll find another wife there. He'll forget us."

Physical expressions of feminine ideals were intended primarily for observation and appreciation by other Somalis, specifically women's husbands and families. Although women were actively trying to make themselves more beautiful, and were more than happy to talk about it and indeed involve me in it, it was not meant for the public, or at least, young women acted and were expected to act as though their beauty remained private. It was common to see young men casually relaxing on street corners, admiring a pretty woman as she walked by with the exclamation of "*Masha'allah*!" [Praise to God], and if they were feeling particularly bold, a comment on a specific aspect of her form. I was told that respectable women will only respond to this with a glare or scolding the men in return with a retort such as "What do you know about my smile/face/hips?!" Although I saw this many times, I also often witnessed women who were otherwise seen as very respectable responding with a smile or some discreet flirting, with the reaction highly dependent on the man making the remark. Similar to Cvajner's (2011) description of immigrant women from the former USSR in Italy, achieving ideals of femininity, or in her case "hyper-femininity," were meant only for family, friends, and the familiar public, rather than the host community (2011, 361). Although they differ enormously on what those ideals are, both the women I met and the women Cvajner describes attached such concepts of femininity to their own notions of morality and worth, while in a context of marginalization as a result of migration. Interestingly, both perceived the women in their host communities as overly masculine in contrast (Cvajner 2011, 363).

CREATING "GOOD" SOMALI WOMEN IN NAIROBI

These foreign female bodies were one of the most conspicuous visible images marking Eastleigh as distinct from the rest of Nairobi. They filled the malls and markets, working and shopping at stalls selling clothes, beauty products, jewelry, food, and *qaad*.[9] Their bodies covered from the head down, either in a *hijab*[10] or *buibui*,[11] worn by both Somali and Kenyan Muslim women, or a bold, colorful *dirac*, and matching scarf, draped down over their head, shoulders and upper body, which was more readily identifiable as being distinctly Somali. Within Eastleigh, clothes that were deemed to conform to Islamic

Figure 6.1 Somali woman in a market in Nairobi.
Photo credit: Lucy Lowe

modesty codes and Somali fashion were ubiquitous. More than just a continuation of practice and fashion from Somalia, young women in particular emphasized the importance of being seen as "good" Somali women, both publicly and privately, and physical appearance contributed significantly to such perceptions. To be seen as becoming "too Kenyan" was to suggest that a woman had lost her beauty and her moral standing.

Although it may be easy to think about globalization as analogous to Westernization or even Americanization, this is a gross oversimplification and rather ethnocentric view of how people, products, and beliefs move and embed themselves in cultures, societies, and nations other than those they emerged from. This was evident in Eastleigh, not least in how women covered their bodies, with styles, fabric, and clothing originating from a range of countries and cultures. The growth in popularity of what might be identified rather broadly as "Islamic clothing" has not been uniform or universal across different Somali communities, not to mention the much larger global Muslim population. Women over the age of about 30, and even more so, significantly older women, noted a considerable shift in how Somali women dress in public. As Rahma, a woman in her fifties who had lived in Kenya for almost twenty years, explained to me,

> In Somalia women didn't cover themselves the way they do now. They wore *dirac* or the other traditional dresses. In Mogadishu some women wore Western clothes, but that was before the war. Now we see these new people arriving from Somalia and they are completely covered, even with gloves and the veil! And they say "this is how we dress because we are Muslims," but I'm a Muslim and I don't dress like that.

Some people attributed this rise in popularity of more "modest" clothing to the increasing influence of political Islam, both within Somalia and elsewhere. The influence of *Al Shabaab*, a violent extremist group that controlled much of Somalia at the time of my fieldwork, was also frequently cited as the cause. However, this fashion was also notably popular among young Somali women still in their teens and early twenties, many of whom had been born in Kenya or had spent most of their lives there. Wearing *buibui*, particularly those imported from Dubai, was a way of accessing goods from "outside" that were modern and fashionable, but also made a very clear visual statement of how one wanted to be identified as a Muslim woman. These young women appeared to straddle diverse global influences when deciding how to present their bodies, as they covered themselves completely in public, yet hidden under their robes it was quite common to come across a tightly fitting pair of jeans. Sumaya, the young woman I introduced at the beginning of this chapter, showed me photographs of herself that were taken before she was married, and told me how differently she dressed then:

> You can see here, I have jeans on and this cute little top—you've never seen me like this before, have you [laughing]?! I liked wearing these clothes. I didn't have many of them, but it was fun to try them on and show them to friends and sometimes let the boys see you. It made me feel like I was modern. But you can't go out on the street like that or people will think that you're a prostitute or some kind of street girl. My husband would kill me if he saw me dressed like this now, but this is how I looked when we met!

Nadra, a younger, unmarried woman who dressed in a similar style to Sumaya's photographs when in private, but appeared even more conservative in public by always wearing the *niqab*, expressed a similar opinion:

> It's very important to find a good husband, and to do that you have to show that you are a good girl. You have to show everyone that you respect your culture and your religion. Boys know they can't mess around with a girl who is dressed properly, so they treat me with respect [...] But you still want to look nice, so if you have the money you make sure that you have the newest styles, have your hair looking good, have henna done...All those things. [And when asked what was so special about clothing from Dubai,] The nice things always come from Arab countries, because that is where our religion came from, but also because they have money there, so they know how to make things look so beautiful.

Yet actively attempting to be a "good Somali woman" in a physical sense was not simply a matter of what one wore on one's body. Women were engaging in rigorous routines in order to make themselves "beautiful." Although this may sound rather superficial, popular concepts of beauty and the importance attached to them were deeply entwined with perceptions of marriage and motherhood, the two aspects my informants considered most central and fundamental to their lives. Deeply embedded within these roles, beauty was highly influential to how women saw themselves and others as Somalis and Muslims. For these women, appearing physically attractive to their husbands or prospective husbands was perceived as an essential part of securing their marriage (vom Bruck 2002), and thus becoming a woman.

Despite access to globalized images of beauty through television and the Internet, popular concepts of beauty among my informants prioritized aspects that maintained ideals regarding Somali ethnic purity, particularly light skin, a straight nose, and soft hair. Beauty was something that my informants widely believed themselves to have a greater understanding and appreciation of than their Kenyan neighbors. It was widely claimed that with their distinct physical features, Somalis were intrinsically more beautiful. As a Somali doctor told me,

Somalis are a very special people. There are not many people who
look like us. Those from Eritrea, Djibouti, some [Ethiopian]
Oromos. We are a special combination of African and Arab. A few
from the north of Sudan and some of the Rwandan Tutsis look
like us, but not many. So you see, it is only a small number.

Although the appreciation of such physical qualities can be
noted in other parts of Africa, including Kenya, among Somalis
they were directly linked with what was described to me as a
"noble ancestry" as well as what was perceived to be their fundamental opposite—Somali Bantus. The word *jareer* meaning
"coarse hair" is an insult, along with telling someone they have
dark skin or a wide nose; to suggest that someone is Bantu is to
say they are both ugly and inferior. This ethnic group has faced
widespread discrimination both within Somalia and as refugees, stemming from the belief that they are the descendants of
slaves and therefore inferior to other Somalis (Besteman 1995).
Through the idealization of what could be described as "non-
Bantu" features, popular concepts of beauty reflect much larger
social practices, particularly the xenophobia directed at Somali
Bantus, and concomitantly Kenyans and other Africans who
bear similar physical characteristics. Beyond that, perceptions of
ideal features also reflect and reinforce widely held perceptions
of how desirable women should be.

On one occasion, I attended a salon near where I lived in order
to have henna applied to my hands and feet. Chairs were arranged
around the room in a circle, facing inwards so that customers were
able to chat to one another. Only a few moments after arriving,
Zamzam, a 20-year-old Somali woman, enthusiastically asked me
if I would also like a face mask or my hair done. I obliged the face
mask, but declined the hair offer, as I was rather unsure about
what it would involve. We started to chat, and she told me that
she did not normally work in the salon, and was only there to
help her aunt who owned it. Her aunt, a rather glamorous-look-
ing woman who I guessed to be in her fifties, meandered in and
out of the salon throughout the four hours I was there, a creamy
orange mask to match my own on her face. As her niece smeared
on the cream she commented that she perhaps should have used a
skin-lightening mask, to which Zamzam snorted, "She's white!"

"Yes, but she's in Africa, and everyone gets darker in Africa," and then turning to me, "you should be careful of that."

Dark skin was perceived as "bad" or "ugly" and therefore women frequently sought to lighten their skin any way they could. This included temporary measures, such as makeup, or more severe lotions and creams that promised to permanently lighten one's skin. The use of skin-bleaching products is common in many parts of the world, and offers a stark reflection of the use of sunbeds and fake tan products among light-skinned people, particularly women, who desire the "beauty" of darker skin over the "ugliness" of pale skin. Like so many beauty products available in Eastleigh, those imported from abroad, very often from India, were seen as more effective than anything produced in Kenya, and significantly more effective than "traditional" powders, which were mixed with water or other creams to form a paste that could be applied to the skin. When I was about to depart for a visit home, one friend asked me to bring back some skin-lightening products for her to try: "The ones over there must be the best, I'm sure they'll work very well." I explained to her that such products available in the United Kingdom, both legally and illegally, were also often imported, and that due to the largely white population and the popularity of tanned skin, there was a greater demand and larger market for skin-darkening rather than lightening products. She looked at me utterly puzzled and suspicious of whether or not I could actually be telling the truth.

This ethnicized perception of beauty was not limited to physical appearance, but also incorporated a concept of olfactory desirability. Scent arose in a multitude of ways during my fieldwork, from the unpleasant stench associated with urban spaces in Eastleigh and other parts of Nairobi, the perceived stench of Kenyans, to the familiar, pleasant, and attractive smells associated with the private Eastleigh behind closed doors, of incense, perfume, and henna. These odors were used to articulate perceived boundaries between Somalis and their neighbors, while also creating and invoking a sense of the familiar. Classen et al. (1994) emphasize the crucial yet often forgotten significance of scent to experience and identity, and how odors can at once be identified with a particular moment, person, or emotion (see also Moeran 2007; Rasmussen 1999). Scent plays a particularly interesting role in perceptions of beauty. Just as particular physical traits can be deemed attractive or repulsive within a particular cultural context, so too are specific scents.

Scent was repeatedly explained to me as one of the many points at which Kenyans and Somalis differed, and interestingly, both would insist that the other smelled intolerable.[12] Somalis informed me that Kenyans were dirty and were either unaware of or actually enjoyed smells that other people found unpleasant, such as strong body odor or the general stench of poor sanitation that lingered in certain parts of Eastleigh and Nairobi more widely. Conversely, Kenyans commented that Somalis wear so much perfume that it is unbearable, as one young woman expressed with a disgusted laugh, "What is that smell, that one they all have?! How come they all smell the same?! My sister says that if a Somali gets on a *matatu*[13] she has to get off immediately because the smell is so strong it gives her a headache."

Two distinct points can be drawn from the beautification perceptions and practices I observed. First, many aspects of "beauty", including light skin, a straight nose, soft hair, and a tall, graceful, body, are things that one can have *naturally*, and are rooted in perceptions of ethnic superiority. Second, some of these are features that women can attempt to improve or attain if they are lacking them, and being seen as aspiring to and trying to achieve higher ideals of feminine beauty was seen as good female behavior, and therefore, beautiful in itself. Skin can be bleached, moisturized, and perfumed, and hair can be relaxed. This tension that women should ideally be naturally beautiful and yet should also strive for beauty in order to attract and satisfy a husband suggests that, beyond "natural" beauty, the pursuit of feminine ideals through attempts to beautify oneself is itself a demonstration of being a good and desirable woman. The act of beautification can be beauty itself (Cvajner 2011, 364).

CIRCUMCISION

Female circumcision may appear to be a peculiar topic to discuss with regard to beauty, as it is so often seen from an outside perspective as something ugly, horrific, and barbaric, all summed up by the term it is often known by—female genital mutilation. Yet for my informants it was quite the opposite. It was an act of beautification. Perhaps this premise seems less foreign when we consider the growing popularity in Western countries of vaginal cosmetic surgery to "improve" the physical appearance of female genitalia, although there are of course significant differences (Essen and Johnsdotter 2004). Furthermore, it was generally the first form of "beautification," or active attempt to create desirable, marriageable women, that a girl experienced, as it was performed on girls as young as four years old.

Although different forms of circumcision are carried out among women in different parts of the world, the vast majority of women in Somalia, and indeed most of my informants, had undergone pharaonic circumcision. Also known as infibulation or World Health Organization (WHO) Type III (WHO 2008, 4), it is the most invasive and involves the removal of the clitoris and the labia majora and minora. The remaining skin is then sewn together, leaving a small hole for urine and menstrual blood, and it is intended to eventually be cut or torn during sexual intercourse when a woman is married. The other form carried out among my informants was termed "sunna," which varied much more, from a slight cutting of the clitoris, to its partial or full removal, and occasionally the removal of the labia minora. These variations put it somewhere between what the WHO terms Type I and II circumcision (WHO 2008: 4). Although pharaonic continues to be the form most commonly practiced in Somalia (Talle 1993; 2007), sunna was often presented to me, particularly by younger people, as a more modern form, more in harmony with Islamic teachings,[14] and therefore growing in popularity.

There is a wealth of literature discussing the reasons for circumcision (Gruenbaum 2001; Hernlund and Shell-Duncan 2007; James and Robertson 2002; Nnaemeka 2005; Rahman and Toubia 2000; Shell-Duncan and Hernlund 2000; Talle 1993; 2007), so I will briefly summarize the reasons my informants gave me for the procedure, whether they agreed with them or not. By removing the most sensitive parts of a woman's genitalia before she reaches puberty, it is argued that her sexual desires

and urges will be curbed, and that she will be able to resist the
temptation to engage in sexual activity before or outside of marriage, thus avoiding bringing shame to herself and her family. It
was also widely noted that sexual intercourse was normally more
painful for pharaonically circumcised women, another factor in
limiting the desire for sexual activity, as one woman explained to
me: "It hurts so much, and I used to bleed every time. I try to
hide it from my husband, because I can tell he doesn't enjoy it
when he can see that I'm in pain." Some women claimed that an
uncircumcised vagina is unattractive, and that the closed, smooth
surface of a pharaonically circumcised woman is both cleaner
and more beautiful. As one older woman, a firm believer in the
benefits of circumcision, explained, "If you don't circumcise a
girl it remains open which causes many problems. It's harder to
keep clean, anything can get in, it smells worse...It's wrong.
And it's ugly when it's left open."

Those with only a very neat, straight scar left behind were
deemed the prettiest of all, and a thick, prominent scar was
perceived as the work of a circumciser lacking in skills. In addition to making her aesthetically pleasing, her sewn-up skin is
intended to be evidence of her virginity and therefore moral
respectability. During a discussion with a group of women in
their early twenties who had all been pharaonically circumcised,
one told me:

> If you're circumcised you have to be careful how you behave
> and move around. You can't run around like you did when you
> were a child. If you jump around like this [she begins to demon-
> strate, jumping up and down with her legs slightly apart], you
> might tear or damage something down there. Then it looks like
> you've done something that you've not! So you make sure you
> walk nicely, like this [she performs an exaggerated graceful sway
> across the room, to the amusement of her friends].

Discussions on the topic of circumcision were rarely straight-
forward or simplistic, and it was not unusual for an informant
to tell me something during one conversation, and something
completely opposite the next time I spoke with her. This was
particularly the case when I asked women who were yet to
have any daughters whether or not they intended to circumcise

them. It was often argued that circumcision was an intrinsic aspect of "Somali culture" and should therefore be preserved and maintained, yet some of the same people who made that argument also told me on other occasions that "Islam does not demand it" or that it was an outdated and "primitive" practice, only carried out by uneducated women. This illustrates that not only were women renegotiating what the practice of circumcision means to them, but they were also reexamining how it fits in with ideas of religious and cultural requirements and norms. Sabrine, a young, unmarried woman who had been circumcised but was opposed to the practice, discussed how Somali "traditions" were changing:

> If you ask some old women they will tell you that pharaonic circumcision is our tradition, but it's called pharaonic because it came from Egypt. It wasn't even ours to begin with! Or they tell you it's religious, but our religion didn't come from the pharaohs, and as we are experiencing more of the world, we are learning that this isn't something we have to do. Other Muslims don't do it. And Somalis living outside don't do it. Some of them do, but not all of them. We are learning that this tradition is bad and we should stop doing [it].

It is evident from these examples that the global spread and exchange of knowledge, as well as goods, has had a profound impact on how women identify what they should and should not do, and in this case, how to properly maintain cultural and religious traditions. While some perceived this exchange as a threat to traditional values, others embraced it as a way to explore new understandings of what it means to be a Somali woman.

CONCLUSION

The depiction of Somali women in mainstream media is most often that of the starving refugee, the mother desperately trying to save her children, the innocent victim of war. Alternatively, they are lumped together with other Muslim women in the vague, orientalist perception of the exotic eyes lurking demurely behind a *niqab*. Although there is some truth to such depictions, as the women I met had been forced to flee their country as refugees, and many had been the victims of violence, these images dehumanize and victimize such women. Rather than looking holistically at personal experience, they are reduced to little more than negative narratives, or rather, the generalized experiences of suffering in Somalia.

It could be argued that focusing on how women beautify their bodies also overly simplifies or objectifies women. However, as I have tried to highlight, beliefs and practices surrounding beautification reflect existing and historical social practices within Somali society, including specific perceptions of gender roles, the emphasis on the importance of marriage and childbearing, and the discrimination and marginalization of particular ethnic groups. In practicing beauty, women were actively trying to preserve a sense of "Somaliness" through their bodies, in the face of new global influences. At the same time, many women saw their bodies as a rare site of agency, through which they could achieve full personhood by creating a desirable but also morally respectable self. In the context of displacement and as part of a vast, sprawling diasporic community, many women were engaging in activities unthinkable to their mothers' generation. They were attending school, and sometimes college and university, some were in paid employment, and many spoke more than one language. In one way or another, either through their consumption of imported goods, their observation of foreign television, movies, music, and fashion, their communication with friends and family in other continents, or even the simple fact they lived in a foreign country, all of my informants were participating in globalized interactions that were in turn influencing and reshaping how they perceived and acted upon their own bodies.

To a certain degree, it is globalization that has been the catalyst for how many Somali women present their bodies to the rest of the world. The emergence of political Islam both in Somalia and in some of its diaspora communities, including Eastleigh, has seen the full covering of female bodies and sometimes faces replacing more "traditional" and less conservative forms of dress. At the same time, exposure to global products means that women use makeup, perfume, and beauty products and practices—imported from a wide range of countries, and often perceived to have special value and capabilities—to help them achieve their ideals of femininity. However, it has also exposed people to foreign aesthetics that they do not desire and actively avoid. Somali women, many of whom see the conflict in their home nation as infinite and inexhaustible, for whom hopes of return are therefore an impossibility, are actively renegotiating how to physically be a "good" Somali, Muslim woman.

NOTES

* This work was supported by the Economic and Social Research Council [grant number ES/H018913/1].

1. This chapter will use Anglicized spellings of Somali names, that is, "Fatuma," rather than "Fadumo" and "Saido," rather than "Saciido." My informants regularly switched between both spellings, depending on the context, and generally recognized both as correct. It is assumed that non-Somali speakers will find Anglicized spellings easier to comprehend, particularly as they correspond to popular Arabic names.

2. This policy was slightly relaxed in 2011, allowing refugees to register with the Department for Refugee Affairs in Nairobi, before being tightened again in 2012.

3. The exact number of Somalis living in Eastleigh is unknown and fluctuates, but has been estimated to be between 40,000 and 100,000.

4. Somali money transfer businesses.

5. The reasons put forward by my informants for hair removal were rooted in *hadiths* that stipulate what hair should or could be trimmed or removed, and what must be left uncut. There was some variation between informants about what could be removed, for example, I heard differing opinions among women on eyebrows and other facial hair, but there was general agreement that most if not all body hair should ideally be removed.

6. A veil worn over the face, leaving only the eyes visible. Only a minority of women wore the *niqab* in Eastleigh, and among my informants it was almost exclusively worn by women under the age of 40.

7. In Eastleigh this term was used to refer to men who had studied the Qur'an extensively and had thorough knowledge of Islam. They were widely recognized as both religious and social leaders, and those who were highly respected held considerable influence within the community.

8. A style of dress that is largely associated with the Indian subcontinent.

9. A plant commonly chewed for its stimulant properties.

10. Although this term is used to describe different things in different global contexts, among my informants it referred to a large garment worn outside the home, designed to cover the head, shoulders, and upper body, leaving only the face visible (*hijab* was used interchangeably with *jilbaab*, however the former was used much more commonly). Like most clothing worn by Somali women in Eastleigh that was intended specifically for wear in public, *hijab* were normally plain in color, and as well as being popular with women, it was also the most common style of school uniform for girls at the privately owned Somali schools in Eastleigh.

11. A full-length, long-sleeved dress worn outside the home to ensure the body is entirely covered, except for the head, hands, and feet. Most often black in color, they are distinctly less colorful than *dirac*, and are normally made of a heavier fabric. They were not, however, immune to fashion, as various styles, including diamante decorations, tailoring, or hooded attachments to cover the head came and went in style.

12. See Classen et al. 1994 and Low 2006 and Manalansan 2006 for further discussion on scent as ethnic identifiers and boundaries.

13. A type of minibus that is the most common form of public transport.

14. Among Muslims the term *sunnah* or *sunna* denotes the practices and teachings of the Prophet Mohammed, and many of my informants argued that the form of circumcision that was termed *sunna* was therefore beneficial. It should be noted that female circumcision is by no means inherently linked to Islam. Most Muslims do not practice it, and it is also practiced, including the form termed *sunna* here, by non-Muslims, including Christians.

7

THE BEFORE-AND-AFTER TEMPLATE

RESEARCHING AND REFLECTING ON BODY IMAGE CONCERNS IN GLOBALIZING INDIA (*Personal Reflection*)

Jaita Talukdar

There was a time when I was quite an admirer of the before-and-after (hereafter BAF) template that applauded the transformation of the body from being fat to becoming thin. The ubiquity of BAF stories to capture the transformation of the body shows us how deep-rooted our distaste has become for not-so-thin bodies, where being fat is always the before story and always a shortcoming. In this essay, I too use a BAF template, albeit one that is stretched out in time, to reflect on my personal struggles with body weight, and my quest for thinness. I come to this story both as a young teenager growing up in urban India in the 1990s, desperate to fit into a smaller size jeans, and as a feminist sociologist who only after many years of researching women's body image concerns has realized that my weight issues were intimately tied to my struggles of expressing who I was or wanted to be in a world that was rapidly changing around me. Hence, a before and an after in this story.

* * *

THE BEFORE STORY: GROWING UP WITH "CHUBBY CHEEKS"

Born and raised in the city of Kolkata (previously Calcutta), India, for the most part I led an upper middle-class lifestyle and enjoyed its many privileges. My parents owned a house and a car, and most importantly sent me to a private, English-language school and invested in my academic career. We were a bit Anglicized as well, which meant that we were comfortable speaking the language and could seamlessly step into and derive pleasure and enjoyment from English-speaking worlds. Some childhood memories of an early induction into this world were my mother reading Mills and Boon romance novels, and enthusiastically sharing stories of the lives and exploits of Hollywood stars like Liz Taylor, Sophia Loren, and Audrey Hepburn's extra-thin waistline in the film *Roman Holiday*. Clearly in awe of the looks of Western actresses, my mother, however, was very comfortable with her own body weight and never aspired to be thin, nor did she transfer any thinness aspirations onto me. Rarely did my parents or my extended family members say something about my appearance. Except for the occasional "chubby cheeks" and "baby fat" used to describe me or my weight, no one made fun of my weight gain. My parents, in fact, constantly warned me of the perils of being underweight at a young age and foiled most of my attempts at dieting or wanting to skip a meal. In the larger scheme of things it was not my appearance, but only good grades in school, that mattered.

Yet, my early exposure to Western norms of thinness did leave its impression on me. As a teenager I started gaining weight that bothered me. In spite of not owning a scale at home, I knew from standing in front of the mirror that my body was changing. As someone with a genetic predisposition to being well-rounded, the weight stayed with me, as did the unhappiness surrounding it. From the many moments of discontentment to choose from, I clearly remember the one that involved going to buy a new pair of jeans with my parents in tow. I had outgrown my old clothes, and this was going to be my new pair of jeans as a young teenager. Before the advent of malls and shopping complexes, folks with good taste in Kolkata, which often meant a penchant for Western goods, shopped at a bazaar called New Market. The shop that we walked into was owned by a man and his male siblings who sold women's clothing. The man took out a couple of pairs of jeans for me to try and showed me the makeshift dressing/fitting room cramped into the corner of the store. My memory of that afternoon is a little blurred except for the cramped room and the moment when, on wearing the jeans, I saw fat on my clothed body that I was unaware existed. The snug fit of the straight-lined jeans had accentuated my body fat and brought it into plain sight.

While now I can look back at that afternoon and have a good laugh, as a young teen, I remember being engulfed in shame and embarrassment on seeing my own body in the neon light of that cramped room. Walking out of the dressing room with my fat exposed for all to see was very unnerving, though that was not the whole story. I took that moment as a personal failure. For me wearing a pair of jeans that showed a slim body was the surest sign of being youthful and modern, in addition perhaps to my ability to understand and communicate with the English-speaking world. I led a very restricted life as a teenager—went to an all-girls' school, was chaperoned everywhere, and my parents had the final say about where I went and who I could spend time with, which generally excluded most forms of male company. Thus, experimenting with new styles of clothes—as long as they did not show too much skin—was one of the few tokens of freedom available to me. Moreover, clothes helped signal to others that I was cognizant and in step with global standards of style. As the country liberalized its economy, Indians were introduced to unknown commodities such as MTV and Fashion TV, which were setting new standards of lifestyles and bodies. The body that seemed to best complement this new world was a thin body just like that of the women featured in fashion channels, television series, and news. It was the yearning to be part of this new, exciting global world that seemed to exacerbate my need to fit into a pair of jeans, especially since what you wore was an important criterion of where you found yourself placed in the traditional-modern binary.

Put simply, your ability to comfortably fit into Western clothes indicated that you were not only eager to, but that you would naturally fit into this new world. There were even derogatory terms for young women who could not fit into this new world such as "*behen-ji*," "*behen-ji turned modern*," or "*geinyo*." These terms were all vernacular, colloquial terms used to mock women's traditional style of clothing. *Behen*, a Hindi term meaning "sister," is generically used to respectfully address older women by adding the suffix *ji* to it (to denote respect). But in school corridors, *behen-ji* was a derogatory term used to refer to young girls who were ostensibly old-fashioned in their ways such as wearing the two-piece *salwaar kameez* (two-piece attire worn with a long scarf draped across the shoulder) or accessorizing with bangles or a *bindi* (forehead art). *Geinyo*, a Bengali term, is a derivative of the word *gram* (village), but was similarly used to refer to women who wore only traditional wear.

No one referred to me as *behen-ji*, but that did not mean that I was comfortable in my non-Western clothes. What followed after the dressing room incident was that I took a hiatus from wearing jeans for many, many years to come. I took refuge in loose-fitting *salwaar kameez* and *kurta* (long, straight-cut shirts) to hide my fat. My exposure to bodies of the Western world convinced me that the only body that befits a modern lifestyle is a slim body. As the garment industry changed around me drastically with the influx of more new styles of clothing, my own clothes were a constant reminder that I was failing on a daily basis to get the modern look right.

*   *   *

THE AFTER STORY: REFLECTIONS AND RECONCILIATION

My after story begins in a different part of the world: in the United States, when I moved there for a PhD in sociology. Of course, I brought all my anxieties about body weight with me now that I was literally going to be part of the West—a Western world that absolutely required Western clothes. This time around I was compelled to buy a couple of pair of jeans even if it bothered me. But like most BAF stories, there was the hope that I could lose weight that came in the form of discovering "fat-free" food at the grocery store and a gym. Excited, I stocked up my refrigerator with fat-free everything. It was not, however, until I started going to the university gym that I actually started losing weight. Discovering the gym was a cultural experience—a large, anonymous space where no eyes were on me and I could join scores of nameless people on treadmills and elliptical machines in the single-minded pursuit of working out. In spite of the mirrored walls and a packed room full of complete strangers, strangely, I was not as anxious about my body weight. I could see the after picture, a much smaller size than the size 10 that I was wearing at that point.

It was in the summer of 2004 that I finally reached my after picture—a size 6 with an almost flat abdomen and my scale showing 119–120 pounds. By now, three years of being in the Western world had changed my understanding of body parts and what to do with them to sustain the after picture—squats and lunges for toned thighs, crunches for a flat abdomen, and "reps" for working on the muscles of my body. Fitting into jeans was no longer a traumatic experience and neither were the dressing rooms cramped. In fact, dressing rooms in this part of the world were large and in secluded areas of stores, with mirrors on the wall that made you look much smaller in size. These would all perhaps count as my after moments in my weight loss efforts if someone were to capture it on camera. What I had not expected was that there was going to be another set of experiences around the same time that would make me gain a whole new perspective about my body.

That summer, when I visited my hometown briefly for a vacation (weighing almost 10 pounds less), I was complimented on my weight loss, on how good I looked, and how much I had changed. The most bizarre correlation made between how I dressed and the new me were the changes assigned to my accent and tone of speaking. One of my mother's friends commented that my ability to converse in English had drastically improved. Others teased that it was a clear sign of my "Americanization" considering that I no longer wore traditional clothes. Coming to America had not caused a drastic change in my cultural tastes in clothes or food; it had only aided the process of weight loss by giving me easy access to a gym. I was particularly astonished by the comment that since I had moved to the United States I had developed the ability to speak good English. My self-perception of who I was as a young teen was at odds with how others remembered me.

I remembered myself speaking very good English and even doing a summer acting and public-speaking workshop in liaison with the Trinity College of England. Was it possible that some or perhaps all of who I was as a teenager was completely buried under my traditional clothes? In many ways, the people around me were echoing what I believed in as well growing up as a teenager. In order to do modernity, one had to get the look right and display a combination of English speaking ability, Western styles of clothing and accessorizing, and of course be educated, ambitious, and knowledgeable about the world. Since I had failed to get the look right, people's memory of my other attributes of being modern and outgoing had diminished over time.

The realization that clothes and body size have become inextricably linked to the identity work of modern Indian women became much clearer as I researched body image and eating concerns of women living in contemporary India as part of my dissertation research. During my fieldwork in 2005, I found many cases of American influences in Kolkata—multiplexes that show recently released Hollywood films, newly built malls that house American franchises such as Tommy Hilfiger, Gold's Gym, and GNC India. The women I interviewed seemed to be implicated in contemporary messages about the body and engaged in different kinds of diets or aspired to wear Western clothes as part of their new, *modern identity* (Talukdar 2012, 113–114). It was also clear that in urban contexts, the understanding of what a modern, embodied experience must entail was heavily reliant on one's choice of clothes and had to be constantly distinguished from what was not modern, and hence traditional (Talukdar and Linders 2013). This is not to suggest that women who dress in Western clothes do not wear traditional clothes; they do. Being modern is, nevertheless, a privileged position because one can be omnivorous in their choice, whereas wearing only traditional clothes implies that you are either failing or not fully engaged in processes of modernity.

My after picture of a size 6 did not stay for long. I quickly gained back most of the weight once I stopped going to the gym, unable to accommodate a routine of working out three or five times a week. Thus, the after picture in my case is not what a BAF story espouses, a body captured and frozen in time. I suspect it is true for others as well. My after template is more of an understanding or a *feeling* that I have arrived at about my body as a result of my pursuits of a thinner frame. It is a feeling of reconciliation and acceptance that I am genetically predisposed to being curvy and that my jeans size will fluctuate between a size 8 and a size 10. Since then, I have also embraced my traditional clothes and the realization that they do not represent a lack of choice on my part to express who I am.

* * *

WHAT NEXT?

The BAF template provides a rather simplistic and compressed account of varied moments of self-reflection strewn in time that underlie not only the transformation of the material body but also, what modernist theorists call struggles of the self. Bodies, however, will always be a tangible and very powerful medium of expressing our self- identity (Turner 1996). After all, bodies play an important role in our social lives—in how we creatively express ourselves, what we wear for private and public occasions, or how we use them for social protest movements. I have not stopped going to the gym or monitoring from time to time what I eat, but none of that is directed toward becoming a much smaller size. As a woman, though, I have come to realize that clothes and body types are used very narrowly to define who we are. Excessively thin bodies are used to represent the modern and successful woman. Being part of a modern, global world, however, is supposed to have the very opposite effect, where there are multiple and varied styles to choose from. My modern identity, thus, now lies in the choices I make about my body weight and the clothes I wear and shielding myself from the temptations of giving into the one-dimensionality of very thin bodies used to represent success and achievement in contemporary times.

8

METROSEXUALITY AS A BODY DISCOURSE

MASCULINITY AND SPORTS STARS IN GLOBAL AND LOCAL CONTEXTS

Jan Wickman and Fredrik Langeland

Nations are socially constructed "imagined communities," as Benedict Anderson (1983) put it. As such they are dependent on symbols that create a sense of unity and community between people who will never interact directly. Human figures and bodies often serve as such symbols, either as allegorical personifications of nations (often female, such as the Statue of Liberty and Marianne of France), or when the exceptional fame of a real-life hero of one kind or another (mostly men) elevates him to the status of an idealized icon of the nation (Gordon 2002; Landes 2001). The strong male body that is a worthy representation of a nation is generally, in the Anglo-American and Western European countries at least, a masculine, hard, and capable body (Hietala 2003; Jeffords 1994).

Anderson (1983) mentions sports competitions, particularly, as examples of symbolic events that serve to generate nationalistic sentiment. Against this backdrop, it is almost self-evident that the athletes who use their able bodies to compete for their nation in international arenas and defend its honor as symbolic warriors will function as mediated embodiments of a nation. This applies particularly to male athletes, as sports still constitute a predominantly masculine arena. The male dominance of sports is reflected in the media coverage. Practically universally, including the Nordic region that is generally reputed to be relatively gender equal, men are the focus of an overwhelming majority of all sports-related reporting.[1] In addition, the nationalistic sentiment that thrives in the sports context is attached more to sportsmen than to female athletes (Agergaard 2004; Bach 2002; Lippe 2002).

An example of the agreement of body ideals and national identities is presented in Susan Jeffords's analysis of the proliferation of a hypermasculine trend in popular culture and politics in the United States during the 1980s. Drawing on classic metaphors of society as a large social body, she claimed that masculine "hard bodies"—in contrast to feminized "soft bodies"—were as collective symbols "widely accepted as the projection of the national body itself" (Jeffords 1994, 26). For the audience, a double identification takes place. First, identification with the individual body, as citizens wish to see themselves as that body; and second, identification with that body as a "collective symbol for a nation that individual citizens receive pleasure from feeling themselves a part of" (1994, 26). Jeffords indicates that this hard, heterosexual, white body has an ideological function in the making of a national ego.

However, in the past few decades the traditional masculine hard bodies and nationalistic ideals of sports have clashed with increasingly eroticized media representations of male athletes. The sexualization of sportsmen is part of a more general eroticization of the male body as an object of visual pleasure. This is a trend that would seem to challenge the conventional gender constellation of men as viewers and women as objects of their controlling gaze (Mulvey 1975). In popular culture as well as in academia this trend has been associated with the concept of *metrosexuality*, which signifies a preoccupation with appearance and a new "looked-at-ness" of urban young men (Coad 2008; Simpson 1994; 2002). In these discussions, the role of star athletes, such as British soccer player David Beckham, as conspicuous icons of metrosexuality has been intriguing because of the challenge that they arguably constitute to traditional gender orders. This seems to be particularly accentuated when encountered in sportsmen, who are often considered the epitome of traditional masculinity.

Ethnicity is important in a discussion of sports as an arena that carries both strong nationalistic and masculine connotations because it is central in the construction of both national identities and masculinity (through sexuality). In this chapter, we will discuss how an analysis of metrosexuality can highlight diverse meanings attached to the sportsman's body in relation to the four intersecting categories of gender, sexuality, ethnicity, and nation(ality). Our interest lies in representations of influential male athletes in the postmodern context, a phenomenon that the British media scholar Garry Whannel (2002) characterizes as *media sport stars.*

Metrosexuality is by now deemed to be an international, even a global phenomenon (Coad 2008). However, it cannot be expected to be identical across different cultures. While contemporary media content and popular culture are disseminated globally, national, regional, or other culturally and socially relevant contexts still play a part in how discourses like metrosexuality are adapted and understood. In order to highlight the significance of a specific local context, we want to discuss the international variation of metrosexuality by contrasting an analysis of examples of sexualized media representations of athletes from the Nordic countries with the Anglo-American literature on metrosexuality. On the one hand, the Nordic countries, the United States, and the United Kingdom are included in what is perceived as the Western cultural sphere, in which there is commonality between countries in the way national identities are constructed through an opposition to non-Western others. On the other hand, there are significant differences in gender cultures and constructions of national identities that constitute different frames of reference for the interpretation of a body discourse such as metrosexuality. By investigating the international phenomenon of metrosexuality in sports with particular attention to the significance of the local context, we thus hope to nuance the discussion of its complex meanings.

We analyze three pairs of images of sport stars from Norway, Finland, and Sweden, respectively. Each one of the pictures is discussed in relation to the overall public personas of the depicted sportsmen. Our intention is not to compare the three countries from which our examples originate. The narrow material might not be sufficient for a nuanced comparison of sports publicity in the three countries, or a discussion of the existing differences in the three national cultural, social, or political contexts. Instead, we think of our chapter as a discussion located in the Nordic region that is made up of nation-states of a similar size, with relatively similar social systems and, perhaps most importantly in this context, parallel gender cultures.[2] We view the images as a series of *articulations* in which both transnational and local cultural norms and attitudes are being expressed in a specific context (Schlack 1996).

How does a sportsman's metrosexuality correspond with different national circumstances? What is the role of ethnic markers of the body in this interplay? How does the special position of athletes as nationalistic icons affect the way their metrosexual appearances in media are viewed? In other words, how is a phenomenon with practically global spread, such as metrosexuality, modified in different national settings?

MASCULINITY AND THE METROSEXUAL BODY

The term metrosexuality refers both to an increased aestheticization of the corporeal expression of masculinity and a sexualization of the male body. However, the concept that became a trendy buzzword in academia and popular culture in the early 2000s is not limited to appearances only. The British columnist Mark Simpson (1994, 2002), who originally introduced the term, included in his concept the heterosexual employment of sexual practices that more traditionally had been associated with gay men. Among these he mentions an interest in the erotic potential of the heterosexual man's rear, anal stimulation, and male passivity. Similarly, David Coad, the author of *The Metrosexual: Gender, Sexuality, and Sport* (2008), states that metrosexuality is not merely about a visual manner of representing the male body, but also involves men who are ostensibly heterosexual taking a relaxed attitude toward same-sex desire. Consequently, metrosexuality becomes a queering practice that diminishes the divide between homo- and heterosexuality. Coad describes it as nothing less than the potential beginning of a "metrosexual revolution" in masculinity and gender relations:

Metrosexuality is replacing traditional and conventional masculinity norms. It may in time become itself the new norm, transforming the way men treat their bodies, how they interact with women, and how they perceive non-normative sexualities. The metrosexual future is one in which men demonstrate more human and more humane values. (Coad 2008, 198)

There are different views among scholars on the significance of the increase in the overtly sexualized images of male bodies as a possible sign or agent of change in masculinity constructions. Several critical voices, such as Swedish social psychologist Thomas Johansson (1998) and British cultural studies scholar Richard Dyer (1992), emphasized that the ideal bodies in this imagery are hard and impenetrable. Accordingly, what we see is simply a "new" expression of a very traditional kind of tough and dominant masculinity. Others, such as the American feminist philosopher Susan Bordo (1999) and British cultural analyst and feminist Rosalind Gill (2009), have put more focus on the undeniable change in the tone of media imagery over the past few decades and discussed the significance of male bodies appearing in the position of an object of a desiring gaze, a position that more traditionally has been associated with submission and femininity. Simpson (1994, 194) eagerly stated that traditional heterosexuality and masculinity "cannot survive this reversal." While Gill recognizes representations of the undressed male body as one of the most profound shifts in the visual culture during the last decades, she also wishes to emphasize the significance of the gender, race/ethnicity, and class differences visible in the processes of the "sexualization of culture."

We would like to draw attention to the dual character of this "new" focus on the appearance of young urban men. It can be said to consist of two aspects: an aesthetic and an erotic (Wickman 2006, 149). The aesthetic refers to an increased interest in fashion and grooming while the erotic refers to the increased tendency to view the male body as a sexualized object. In the latter case, metrosexuality is often less about fashionable clothes than a fashionable lack of clothes. While these two aspects are admittedly intertwined, we would argue that the distinction between them is significant. Breaking appearance-related, gender-related, and sexual norms may have different consequences in different contexts. However, even if sexually charged imagery today is almost omnipresent in Western media and public spaces, sexuality is still a provocative issue in many contexts and saddled with many taboos. Sexuality is also—like gender and in connection to gender—perceived as a deeply personal and important part of an individual's personality. In contrast, appearance-related practices such as aesthetic metrosexuality can be perceived as something rather superficial and disregarded if they are not given sexual meaning and thus viewed as a (potential) reflection of identity and/or a challenge to heteronormativity. Therefore, it could be argued that erotic metrosexuality is in many contexts more provocative than aesthetic metrosexuality.

The distinction between the aesthetic and erotic aspects of metrosexuality is significant, for example, when reading Coad's aforementioned book (2008), which is the only academic book-length study on the concept of metrosexuality to date. Coad arrives at his optimistic assessment of metrosexuality after a discussion in which he seems to put focus mainly on the aesthetic aspect of metrosexuality when discussing how widespread the phenomenon has become, and on the erotic aspect when discussing its meaning and significance in relation to masculinity. Furthermore, while his wide overview indicates that metrosexuality has become international, bordering on global, noticeable international variation in the emphasis of its two aspects begs for further analysis. The large selection of cases of metrosexual sportsmen from many countries suggests that the public erotic aspect of metrosexuality gets more emphasized in Europe and in some Latin American contexts than in the United States, where it, in addition, remains largely concentrated on sportsmen of color (see Wickman 2011).

Sportsmen of Color Representing White Nations

In the European and Anglo-American visual culture, hegemonic white masculinity has been normative and unmarked. As such, it represents a dominant power position while the black body is often explicitly marked and subordinated in the public space

(cf. Dyer 1997). In accord with widespread racist and stereotypical perceptions, black men have been portrayed as especially
erotic and sexually active, and thus subject to an objectifying
gaze (Carrington 2002, 3; Gullestad 2007, 181; Hall 1996).
In a society that traditionally has defined hegemonic masculinity as strength/activity and femininity as weakness/passivity, it
will arguably be problematic for a male body to be subjected to
such a gaze (Mulvey 1975; van Zoonen 1994, 98). British sociologist Kobeena Mercer (1994) argues, for example, that the
fixation on the racialized black body maintains the white male
ego-control and the colonial fantasy of the other. Indeed, also
in contemporary media sports, the sexualized black body has
been considered in many ways reminiscent of the sexualization
of the female body (Carrington 2002).

Whiteness constitutes a significant element of national identities both in Europe and America, although in different ways. In
European nation-states the dominant white groups claim special legitimacy in the capacity of age-old inhabitants of the land.
This is accentuated in the Nordic countries that, unlike larger
and more politically powerful European nations, do not have
extensive histories as colonial powers that would have required
legitimization in overtly racist terms. Instead, their relation to
colonialism has been described as characterized by complicity
(Tuori et al. 2010). Such traditional definitions of the nation as
white that defy new realities of societies in a globalizing world
are countered by a postmodern view of a multiethnic nation.
For example, the constructed image of the Nordic countries
as ethnically homogenous and monocultural is challenged by
new national self-images, characterized by multiculturalism, as
a result of the country's existence in an international community (Hedetoft 2003; Mühleisen and Røthing 2009).

These different ways in which whiteness takes a central role
in the construction of a nation have consequences for how athletes of color are seen as representatives of that nation. In the
United States, many of the most successful athletes in several
prominent sports such as football, basketball, or—to mention a
sport in which US athletes are visible representing their nation
in international arenas such as the Olympics—track and field,
have long been African American. Other sports have remained

white for much longer. Similarly, several of the great stars in the British Olympic teams have been men of color while, for example, elite teams in soccer (a major sport in the country) have become more multiethnic relatively recently. In the Nordic countries, the star athlete of color exists, but is still often more of an exception. For example in Finland, Northern or Nordic whiteness (with mental features—such as cool calculation—assumed to be associated with it) was, in the early twentieth century, one of the central national characteristics attributed to Finnish star athletes in media coverage (Tervo 2002). Even later, at a time when the nation is more multicultural and multiethnic, a perceived lack of what is considered authentic Finnishness has diminished the star and hero treatment in media of athletes who have gained considerable success in international sports events competing for Finland (Kinnunen 2003). Similarly, in Norway, sportsmen of color have traditionally not been perceived as authentic national heroes (Andersson 2004).

Important elements to take into consideration when discussing visual culture in the Nordic countries are contemporary national self-perceptions of gender equality. In international comparisons, the Nordic countries have in the past decades led international rankings in gender equality statistics with regard to such fields as economic participation, educational attainment, health and political participation (Hausmann, Tyson, and Zahidi 2010). Gender equality is also deeply embedded in the legislation and the welfare state systems of these countries (Kjeldstad 2001). While it is agreed that gender equality problems are by no means solved in the Nordic countries and feminist critique is still pertinent, pride over the achievements in this sector has become a part of the national self-image. In Sweden and Norway, there has even occasionally been self-confident talk of exporting these countries' gender-equality models and policies. The national self-confidence in gender-equality issues is reflected also in self-righteous public debates that label immigrant cultures as patriarchal and as having particular equality problems (Mulinari 2008; Skille 2006; Strandbu 2005). The Nordic societies, particularly Sweden and Denmark, also have a reputation of being liberal and permissive in issues regarding sexuality (Arnberg 2009; Plummer 1998). This discourse too has become a part of

Nordic national self-conceptions (Røthing and Svendsen 2009). But the question here is whether the Nordic culture and politics that combine an ostensible emphasis on the dismantling of gender hierarchies with sexual open-mindedness make a particularly susceptible ground for media metrosexuality.

HARRI HAATAINEN AND MARKUS PÖYHÖNEN: ONE SEXY (ETHNIC) FINN MORE FINNISH THAN THE OTHER?

Javelin-thrower Harri Haatainen and sprinter Markus Pöyhönen are the two celebrity athletes in Finland who in the early years of the 2000s got the most media attention on their looks and sexiness. Both of them repeatedly posed shirtless for advertisements and publicity shots. We have chosen two examples that were originally published in advertising campaigns for hair products and sports gear respectively. In them, Haatainen[3] and Pöyhönen[4] pose shirtless in semi-close ups with their hands lifted up to or behind their heads. The similarities in the pose and the crop help to highlight the differences between the images that make them interesting because they represent well the difference in the general public image of Haatainen and Pöyhönen at the time.[5]

Both Haatainen and Pöyhönen are ethnically Finnish, but characteristics that are stereotypically associated with Finnish nationality are clearly more emphasized in the media image of the blond Pöyhönen. For example, in an article in the popular tabloid newspaper *Ilta-Sanomat* (April 18, 2003, 36), reporting that the readers had selected Pöyhönen as the sexiest man in the country, in addition to his Adonis body, he was praised for being a-boy-next-door-type and a "true Finn with a linen hair."[6] He has also repeatedly been called an *Elovena boy*, for example, in a more recent listing of sexiest male athletes of all time in the competing newspaper *Iltalehti*.[7] This designation is a benignly humorous gender inversion of the figure of the *Elovena girl*, an iconic representation of the ethnic stereotype of a Finnish female in an authentically Finnish rural setting. Elovena is the leading brand of oatmeal in Finland, the classic packaging of which has since the 1950s been adorned by a drawing of a young blond maiden in national costume against the background of blue skies and an oat field.[8] Although Pöyhönen is born and bred in the Helsinki metropolitan area, the allusion to his blondness by reference to this nationalistically charged figure connects him with the countryside, which is associated with innocence and is rich in nationalistic symbolic meaning.

In contrast to Pöyhönen, Haatainen moved to the capital from a smaller municipality, but his new urban environment became an essential part of his public image. His taste for the in-crowd nightlife and fashion were recurrent features of the coverage that associated him with the city, which, in the Finnish and Nordic context, is a setting with more cosmopolitan flair than national character. The frequently eroticized representation of Haatainen, was thus associated with features of urban aesthetic metrosexuality that distance him from the traditional image of Finland just as much as they detach him from stereotypical traditional masculinity.

The image of Haatainen (referenced in endnote #3) was originally published in a calendar produced as a Finnish PR item for an international brand of hair products in 2002.[9] This photo constitutes an example of how the imagery of sexually charged representations of men has become more brazen in the past few decades. Since there is a rather scanty tradition of such images of men in mainstream media from the decades before the era of metrosexuality, effects and iconography have been borrowed from earlier erotic images of women and from gay subcultures (Rohlinger 2002; Wickman 2004). Similar to the way we have described metrosexuality itself, these visual sources are internationally, often globally, disseminated, but nevertheless occasionally exhibit national imprints (Adam, Duyvendak, and Krouwel 1999; Altman 1997).

In the black-and-white picture, Haatainen lies on his back on a piece of folded fabric in an arrangement that brings to mind an iconic naked picture of Marilyn Monroe lying on red velvet. His pose is familiar from numerous photographic images of women: his hands rest on the fabric, one is placed gracefully beside his forehead and the other over his head, leaving his pectoral muscles open to view. There is nothing subtle about the way this body is offered as an erotic visual pleasure, as he looks upwards to the right of the spectator. In an analysis of eroticized images of men, Richard Dyer (1992) notes that looking up past the camera/spectator was common in earlier facial pin-up pictures of mainly American male film stars. He interprets it as a sign of the man in the picture fixing his attention to something higher and more important than the viewer, thus refusing contact and availability via the camera lens.

However, in this picture, with the man lying down, naked, this look results in a different effect. It seems like Haatainen gives up control and his body is left in the power of the viewer. In the older American and British male erotica that Dyer also analyzed, the men were posing with their bodies strained as if ready for action. In the picture of Haatainen the muscles are the only sign left of the body's capability for action, as they are combined with an artificially relaxed passivity of the pose.

This particular image displays several means of eroticization that set the imagery of the past 15–20 years apart from the representation of the male body in mainstream media in the earlier parts of the twentieth century. The cropping of the picture brings the viewer closer to the object than a full body image that would more effectively underline the physical capability of the body and distance it from the eroticizing gaze as earlier pictures tended to do (cf. Dyer 1997). Thus the intimacy impacts the impression the image makes. And the viewer's attention focuses on the surface of the skin as some moisture or oil makes it gleam in the light. This is an effect that has frequently been used in older imagery to accentuate the forms of a muscular body; but in this image the result is more sensual than in traditional pictures. Here it entices the viewer to imagine touching the skin, particularly when the picture is reproduced in large size.

The picture of Pöyhönen's upper torso (referenced in endnote #4) was used to promote an international brand of sportswear. His pose with hands behind the head gives an opportunity to show off particularly the abdominal muscles (and body builders use it for this purpose). At the same time, when read as a gesture, the manifest absence of any sign of defensiveness in this posture suggests a both open and trustful attitude toward the camera/viewer, which has made it a staple in an international visual vocabulary of eroticized images of males (Wickman 2004). In this picture, the impression of openness is reinforced by the athlete's gentle look right into the camera with a trace of a smile on the slightly pouted lips. Facial expressions like this could, according to Dyer's (1992) analysis of Anglo-American material, earlier be found exclusively in eroticized images of women. Still, while the attitude and pose bring the athlete (emotionally) close to the viewer, the photographic style has a more distancing effect. In contrast to the almost tactile impression of the skin in the Haatainen picture, here the dramatic and stark sidelight from the viewer's right casts strong shadows that create a cooler mood. While they highlight some of the muscled forms of the torso, they also render large parts of the body surface in the picture, including almost half of the face, practically invisible. The display of the body is effective and the sexual elements are visible in this photo of Pöyhönen but, in comparison with the far more flamboyant one of Haatainen, it seems almost austere.

The difference in these two images is typical of the eroticizing press pictures of these two athletes more generally. The consequences and interpretations of these representations were also different for the public perception of the two athletes, and the images were related differently to their personalities and sexuality. In the eroticized images Pöyhönen was seen as "naturally" desirable. His body shape and sexiness would be regarded as something of a by-product of his athletic career, even if he, in fact, is a good example of the contemporary "aesthetic athlete" who invests in his appearance (Kinnunen 2003). The bulky musculature of Pöyhönen's upper torso hardly enhances his ability to run fast. Still, the visual impression of the images leaves him innocent. Pöyhönen is the subject matter of the images but he is not seen to actively participate in his own sexualization. On the surface, his beauty seems almost accidental. Therefore, most importantly, the sexualizing images do not become parts or reflections of his personal sexuality and they do not question his basic heterosexual manliness.

In contrast, the obvious and deliberate sexualization in the photographs of Haatainen became a central part of his public persona. The effect was reinforced by the fact that his erotically branded appearances were complemented by aesthetic metrosexuality. Haatainen's penchant for fashion was another standard ingredient in media portraits of him. The fact that his metrosexuality was constantly commented on contributed to an impression that it was a part of his personality as well as the basis for his fame. In other words, the other features of Haatainen's public presentation did not contribute to dispel an unspoken doubt about his sexuality, which is raised by his submission, in pictures of the kind discussed here, to an eroticizing gaze that is potentially homosexual.

There is a difference in how apparently heterosexual the two men appear also in textual representations. At the time, in the coverage on Pöyhönen a steady girlfriend and a Christian worldview were repeatedly mentioned. This created an impression of him as more certainly a regular straight guy. In the case of Haatainen, no steady relationship was reported, which would have rendered his personality some sound respectability to balance his reputation as a party boy that associated him with the decadent urban nightlife. Neither was his persona characterized by the kind of brutal manliness that establishes a hetero-masculine tone in the public image of some athletes. It is hardly coincidental that the more pronounced forms of metrosexualization that render the athlete's normative heterosexuality insecure are to be found in the public image of the sportsman (Haatainen) that to a lesser extent than the other (Pöyhönen) is associated with Finnishness.

KRISTIAN KJELLING AND JOHN CAREW: ETHNICITY OVER NATIONAL ATTRIBUTES

The two examples from Norway are from the years 2005 and 2006. They show two male athletes who at that time were on their way to becoming international sport stars: handball player Kristian Kjelling and soccer player John Carew. Kjelling has arguably been one of the most visible Norwegian male athletes in the 2000s—as a successful professional athlete, a national hero, and a model in a variety of commercials. Soccer player John Carew emerged as an important and interesting character in the public in the mid-2000s. He has had a highly successful international career in England, France, and Turkey. Both men are famous also for their interest in clothes and fashion.

Kjelling's position as a prominent figure in Norwegian media was established in 2005, when the Norwegian national handball team had their best tournament ever in the World Championships.[10] This was also the year when his appearance in the media evolved from being just an athlete to becoming a media sport star. During the tournament he received massive attention. On the cover of Norwegian tabloid *Dagbladet,* he was depicted semi-naked with a desirable and objectified body (von der Lippe 2010). In an interview in the newspaper *Verdens Gang* (January 19, 2006) he was later portrayed as a "trendsetter" in relation to his interest in fashion and clothing, and it is commented that he is a "brand" that has just started to develop. The article also states that Kjelling follows in the footsteps of David Beckham—thus explicitly casting him in the role of a metrosexual athlete.

The image of Kjelling that we analyze here (figure 8.1) is taken from a campaign he did for the tanning salon company *Brun og Blid* in 2006.[11] This is a color photograph where Kjelling poses bare-chested wearing a white knitted cap, white underwear, and brown pants. The picture is cropped so that the upper part of the brown designer pants is visible, with suspenders hanging down from the two buttons. The trousers are pulled low enough to reveal the designer label *Dolce & Gabbana* of his underwear. He wears a long necklace with a crucifix. There is a tattoo on

Figure 8.1 Kristian Kjelling.

the right forearm. He balances a long object that looks like a pipe on his neck, arms outstretched to the sides in a crucifixion pose. He stands with his back to a worn red brick wall and his gaze directed at the viewer. The muscles on the upper arms and stomach are clearly displayed, the body nearly hairless.

This photograph replicates all the major elements of an iconic picture of David Beckham that appeared on the cover of the magazine *GQ* in June 2002. Like Kjelling, Beckham poses bare-chested, in a similar Christ pose, a crucifix around his neck, and trousers lowered to show his underwear. In the picture of Beckham, launched shortly before the World Cup, the red cross of the English flag constitutes the backdrop. The image of Kjelling, however, does not exhibit any details that in any sense can be considered particularly Norwegian. The metrosexual athlete seems to be depicted here without any nationalistic connotations. His body functions as a marketing tool in a commercial context. His headgear, tattoo, designer pants, and underwear, also reflect the global spreading of what David Coad characterizes as a "black expressive style" with "transcultural appeal" appropriated by white metrosexual athletes (2008, 131). This connotes hypermasculine stereotypes normally associated with a gangster street style, and gives him an edgy look that plays with racialized images of the black body. However, he has been explicitly marked as a heterosexual athlete with a working-class background, thereby representing a masculinity which in the Norwegian context has a particular, even hegemonic, legitimacy (Langeland 2011).

With a mother from Norway and a father from Gambia, John Carew challenged the stereotypical image of the Norwegian sports hero.[12] Here we focus on an interview titled "The Gladiator" from the tabloid *Dagbladet* in 2005.[13] It is situated in a Turkish bath in Istanbul, where Carew at the time played for the Turkish team Besiktas. The contrast between the non-white Carew and stereotypical Norwegian-ness is emphasized in the text: "John Carew is not fair, and light skinned […]. He is, in a double sense, tall and dark." His temperament and behavior is implied to be different from Northern European norms of restrained self-control. The interviewer points out that Carew has in many ways been a controversial figure in Norway, but he is quoted saying that he really feels he has behaved nicely, even though there had been much fuss in the media about some "fights and scandals" in his career.[14] Carew argues that the attention to these is a Norwegian phenomenon, since "brawls between teammates are quite common outside Norway," due to masculinity norms in soccer. But the journalist describes Carew—born in 1979—as not an adult yet, and for "a young guy, who also is a bachelor," this kind of behavior is deemed quite normal. Carew is portrayed as a "boy," an indication that he still has not quite developed an adult identity, and that the mistakes he makes are just boyish misdemeanors.

In one of the pictures[15] that illustrate the interview, Carew poses upright, leaning against a column. A light blue towel covers the upper body, while parts of the chest are exposed. In another picture he is lying on his back in the foreground. He is wearing shorts with the blue towel still covering the upper body. In the background we see three men being massaged by other men; none of them can be seen to wear clothes. In both these images Carew's body is explicitly in focus. The sexual potential is reinforced through verbal descriptions in the text: "When the photographer takes the pictures John Carew flirts with the camera. He is [...] remarkably photogenic, but actually a bit embarrassed about his attractiveness. So a lens cannot simply be shoved in Carew's face thinking that he loves it. The boy is shy too." In this description we get a characterization of Carew as "photogenic," "embarrassed," and "shy." Carew is not only a "boy" who poses for the camera, but also feminized and eroticized.

He is placed in an environment where the men massaging each other in the background may be associated with potentially homoerotic fantasies, which create distance from soccer's traditional homosocial and heteronormative sphere (cf. Messner and Sabo 1990). Furthermore, the exotic quality of his black (semi) African body interrelates with the exotic qualities of the Oriental setting. This can be related to Edward Said's (1978) discussion of the stimulating and sexual imagination of the Western objectifying gaze that represents a position of power. In the relationship between Orientals and western Orientalists, passivity is the supposed role for the Oriental, while for the Orientalist it is the ability to observe and study. It is as if Carew's blackness itself acts as a license that enables the media to reproduce images of both black and of "Oriental" sexuality. In this case, it is *Dagbladet*'s journalist who explicitly gazes upon the body of Carew, and readers and viewers are invited to join this process. The text describes how the photographer and the journalist blatantly stare at a partially undressed, but also partially clothed, Carew. The erotic potential here is not only in the representation of the black body, but also in the prior expectation that he might be undressed. The caption reinforces the prospect of nudity when the emphasis is on how he is "taking the clothes off his body." But the fact that he, after all, is partially dressed and yet covered in the pictures, reinforces the impression of his ambivalent attire. None of the other men in the photograph have covered the body like Carew. It would be more consistent if he had been undressed, since he is in a Turkish bath, but for some reason he is not.

A set of ambiguities is characteristic of the construction of Carew as the other as he is retained between being Norwegian/non-Norwegian, dressed/stripped, and hyper-sexualized/feminized. Even if the representation of Carew breaks with what often constitutes the idealized Norwegian masculine hard body in sports, it would be too simplistic to unequivocally classify his as a subordinate masculinity (Connell 1995), merely because he was represented as eroticized, sexualized, and thus potentially "feminized." In a period in which an increasingly objectified and sexualized portrayal of male bodies has been observable, it could be said that Carew, as a media sport star, has complicated the stereotypical image of the Norwegian white sports hero. Similarly, in the case of Kjelling, there is a tension between conventionally incompatible characterizations in the representation of him as both a postmodern sexualized media athlete and as an authentic working-class family man. The visual images of him are explicitly erotic, almost queer, but the textual portrayals often downplay the sexuality in the representation (compare to Edwards 1997). Kjelling's white, working-class body is thus predominantly reproducing the aesthetic sides of metrosexuality, while the dark-skinned Carew is more explicitly eroticized.

FREDRIK LJUNGBERG AND ZLATAN IBRAHIMOVIC: INVERSION OF A CLASSIC CONSTELLATION?

The two most famous Swedish soccer players in the 2000s, Fredrik Ljungberg and Zlatan Ibrahimovic, constitute our third pair. Both of them are also internationally well-known media stars playing for club teams in England, Spain, France, and Italy. In this sense they are a part of what Garry Whannel calls a "sporting star system," in which some athletes become particularly influential as role models in the public eye (Whannel 2002, 7). It is therefore highly relevant to look at what kind of discourses in relation to masculinity, sexuality, race, and the body are being articulated in the representation of these two athletes.

In this pair, the connection between ethnicity and role as a sexualized object seems to be the opposite of our previous cases. Overtly eroticized imagery and media representations are a more prominent component in the public image of Ljungberg, who has a Swedish name and is generally considered to belong to the traditional demographic makeup of the nation, than in the case of Ibrahimovic whose immigrant background is an important element of his public profile. With parents from Bosnia-Herzegovina, Ibrahimovic has an iconic status as the lad who has achieved spectacular success despite a disadvantageous background in Rosengård, one of the most notorious suburbs labeled as immigrant problem areas in the city of Malmö. In racist circles he is certainly considered ethnically not Swedish, particularly as his father is Muslim (although Zlatan's own religious views have not been publicized), while mainstream media refers to this rejection of the popular athlete as an example of the unreasonable attitudes of the ultra-nationalists (Gellert 2010).

Ljungberg has been cited in international publications as a prime example of metrosexuality. This is because of his posing in remarkably sexualized pictures, while showing both a keen interest in fashion and an open-minded attitude toward sexual diversity, expressed for example, in interviews and photo shoots for international gay media, much in the likeness of the undisputable pioneer of metrosexuality, David Beckham (Coad 2008, 100–105). The highlight of his illustrious pin-up career is an advertising campaign for Calvin Klein in 2003. Calvin Klein's advertisements have a reputation for the most effective exploitation of the spectacle of an eroticized male body, and the brand is seen as a pioneer in using such imagery (Bordo 1999). This company's collaboration with a prominent athlete attracted considerable attention, particularly in his home country. When the news of Ljungberg's contract with Calvin Klein Underwear was released in Sweden in 2003, it was published on the front pages of the most prominent daily newspapers (*Aftonbladet* and *Svenska Dagbladet*), giving the images publicity before the extravagant campaign even began. Ljungberg's later Calvin Klein campaigns have added to a wealth of emphatically sexy images that have been eagerly reproduced, for example, in fan sites of male pin-ups on the Internet. The particular image of Ljungberg we analyze is iconic.[16] The tone of sexualization in this photograph is different, for instance, from the Haatainen picture that we discussed in the first case. Here virility is emphasized by all possible means. Ljungberg stands in front of a distressed wall with something scribbled on it, which suggests a rough environment. The body is strong, although the musculature is not as developed as Haatainen's or Pöyhönen's. Indeed, Ljungberg looks as if he is trained to be fit rather than to look astonishing.

The photo is cropped just below the hip to include the underpants that are the product to be promoted. All the opportunities to add sexual charge that this cropping offers have been utilized effectively by employing different means to focus attention to the penis that is covered but discernible as a bulge in the underpants, which are designed to emphasize it in the first place. The light that falls on the white pants slightly from the side underlines effectively the impressive volume of their content. Ljungberg has pushed his thumbs under the lining, pulling it down slightly in a gesture that teases the viewer with the suggestion of imminent stripping. The phallic puma tattoo that springs on his loin from where the penis is, contributes to the emphasis on virility.[17] Ljungberg looks into the camera and into the eyes of the viewer, but not in the soft way seen earlier in the picture of Pöyhönen. But neither is this look the stern and challenging gaze that characterizes what Bordo (1999) calls *face-off masculinity*, which used to be more typical in this kind of images of men in the 1990s. Ljungberg's head is ever so slightly tilted to the side and forward/downward and there is a slight hint of smile on his lips.

There used to be few pictures of Zlatan Ibrahimovic deliberately posing shirtless for the camera. Now, some years later, the situation has changed. Although in the early 2000s he was not a typical example of a metrosexual athlete, his sex appeal was clearly recognized. The most typical shirtless images of him however, were shots from holidays in the sun or, like our example, from the soccer field.

In the photograph we discuss here,[18] Ibrahimovic wears sports shorts, with the shirt in his right hand suggesting a situation after a game or a training session. He looks down with a discontented expression. This does not express any relation to the camera/viewer. He is distant and indifferent, as if the picture had been taken using a telescope lens. His sports shorts are low in accordance to current fashion, leaving the lining of the underpants visible. This is also so low that the picture ends up being fairly revealing even if the shorts are far from scanty. Images of this kind often do display the body effectively, and they are indeed (re)used in contexts that are focused on the appreciation of male beauty, but they nevertheless lack many of the effects that (are intended to) enhance the sexual charge of the studio photographs of the kind discussed before. The athlete himself certainly seems to contribute to the creation of an erotic thrill even less than in the case of Pöyhönen that was discussed earlier.

In the example of Ljungberg and Ibrahimovic, the tendency to eroticize the exotic male that we have noted in the previous cases seems to be reversed as the ethnically Swedish athlete is the one that is metrosexualized more intensely. The question arises of whether this should be interpreted as a sign of a normalization and destigmatization of metrosexuality, or if there are circumstances that make this case particular. We have already suggested that the gender-egalitarian and sexually permissive ideals in Nordic cultures might make these societies particularly open to metrosexuality. However, it is also to be noted that we deal here with two soccer players, and many of the metrosexual idols, interestingly, seem to be found in this sport despite its well-known stark and homophobic masculinity norms. This might be related to the fact that soccer is one of the most significant transnational contemporary sports, which complicates the player's position as national icons. The most successful soccer players, such as Ljungberg and Ibrahimovic, spend most of their time playing long cup seasons for clubs abroad. Only occasionally are they summoned to the national team to represent their country for a few weeks in an international tournament. Even when playing for a foreign club team, there is obviously special interest in an athlete in his home country. Still, this constellation modifies the association of the athlete and his body with the nation (see Andersson 2004). Thus, the metrosexuality of these athletes does not clash as intensely with nationalistic sentiments as the case would be for sportsmen who more clearly are viewed as representatives of their country. Perhaps the eroticization of the white(r) sportsman seems less provocative when his career does not unequivocally link him to the nation.

CONCLUDING REMARKS

The selection of cases discussed in this chapter is strategic in the sense that it is meant to bring forth different aspects of the entangled connections between gender, sexuality, nationality, and ethnicity in discourses and representations of masculine bodies. Our focus is on the comparison between the two images in each pair, while the three cases are not entirely equivalent to each other. In addition to the geographic locality, the dates of the images and articles that have been discussed here are significant. They are from the period 2002–2006, when metrosexuality had become a titillating favorite topic in popular culture, because of both its sexual allusions and potentially iconoclastic character with regard to traditional masculinity discourse. The great public attention that these representations of male athletes have attained bear witness to the tensions involved in metrosexuality—particularly in the sports context with its masculine and nationalistic connotations.

The first case, which dealt with the media representations of javelin thrower Harri Haatainen and sprinter Markus Pöyhönen, implies that it is sensitive to mix metrosexuality with the nationalistic sentiment that is prominent in sports publicity, particularly when it involves a potential breach with orthodox heteronormativity. Because of this tension, there seems to be a tendency to distance the most nationalistically charged athlete figures from the most flagrant, potentially emasculating, forms of sexualization in media representation. Both men are ethnically Finnish, but Pöyhönen corresponds more to stereotypes of a Finnish appearance. Consequently, he has, to a greater extent than Haatainen, been represented in ways that associate him with national(istic) imagery. It is hardly coincidental that the athlete to whom national characteristics are more readily associated is the one whose sexualization takes somewhat more discreet and cautious, even innocent, forms. Furthermore, this case suggests that it is significant that, although these athletes were successful in their sports on the national level, neither of them ever won medals in major international competitions; the nation's greatest expectations of triumphs and glory in the international arena thus never rested on their shoulders (Wickman 2006, 150).

The second pair consisted of two metrosexual athletes from Norway, Kristian Kjelling and John Carew. They were prominent representatives of two team sports that have a strong position in the national imaginary: handball and soccer, respectively. In other words, compared with the Finns, they were potentially of a greater significance in the national sports world. Whereas the public image of neither Kjelling nor Carew displayed prominent nationalistic elements, their ethnic difference seemed significant in a manner similar to the Finnish case, but more accentuated. In the images that we discuss, both are represented in a sexualizing manner, but the texts of articles that represent them treat the two very differently. Only in the case of Carew (who, with a Gambian father, is a person of color) do the texts reinforce the erotic potential in the image. Carew is made both erotic and exotic with orientalizing elements, while no textual sexualization takes place in the case of the ethnically all-Norwegian Kjelling. Thus, the aesthetic—not the erotic—aspect of metrosexuality dominates the public image of Kjelling.

The case of the Swedish soccer players Fredrik Ljungberg and Zlatan Ibrahimovic would seem to overturn the picture that we have sketched. In the first two cases, metrosexual trends get mitigated by nationalistic discourses of ethnicity, with some restraint shown in the sexualized representation of those white athletes to whom nationalistic feelings are most keenly attached. However, in this case, Ljungberg, white and with a Swedish name, is the one with a long-standing, high metrosexual profile including a documented interest in fashion, appearances in spectacular advertisement campaigns with an unmistakably sexualizing character, as well as reportedly metrosexual attitudes. In contrast,

Ibrahimovic (who has an immigrant background) was not in the early years of the 2000s, the years of the prodigious introduction of metrosexuality, subjected to much attention on his looks. However, over time, his posing in top fashion ads and shirtless studio photos with a clear erotic emphasis have increased significantly. In 2011, Ibrahimovic was even used as the foremost illustration in an article in one of the leading Swedish papers, *Dagens Nyheter*, on men as sex objects (Letmark 2011). This reflects the more general trend of metrosexual representation becoming an increasingly commonplace part of male star athletes' public images (Wickman 2006, 154). However, consider also that soccer as a globalized sport in which the nationalistic ties are often suspended today, combined with the fact that the Nordic region represents a cultural context with gender-egalitarian and sexually permissive ideals, may constitute particularly conducive contexts for metrosexuality in this case.

Gill (2009) emphasizes that the "sexualization of culture" is a heteronormative, gendered, racialized, classed, and ageist phenomenon. Still, referring specifically to sexualized representations of the male body in advertising, she also notes that it is an ongoing, dynamic process, in which imageries change as previous barriers are broken. This change does, however, by no means constitute unequivocal progress toward nonhierarchical gender relations. A combination of commercial, political, and cultural factors facilitates the proliferation of sexualized images of male bodies, as advertisers, aware of feminist critiques and increased influence of gay movements, seek to attract a wide audience. Advertising responds to feminist and queer critiques but does so following its own capitalist logic and commercial prerequisites (Wickman 2004). The seemingly paradoxical coexistence of conservative and more iconoclastic features in our material reflects this continuing state of change and its restraints.

A great majority of Coad's (2008) examples of white athletes being represented in mainstream media in an unmistakably sexualizing manner are from Europe or Latin America, while the most conspicuous American cases tend to present black athletes (or photo models who are not known as sportsmen). In the Nordic context, we ask if metrosexuality is less threatening to white men in a gender-equal discourse, implying that a feminized position is viewed as less problematic. Or is the metro-sexualization of white sportsmen simply a result of lesser ethnic diversity in the Nordic region than in the United States, so that the group of media sport stars is predominantly white? At the same time, the tendency to direct the objectifying gaze toward ethnic others is by no means absent in our material, although an earlier longitudinal analysis suggests that it seems to be decreasing somewhat over time, at least in Finland (Kinnunen and Wickman 2006, 176; Wickman 2006, 154–155). The same discursive structures that are apparent in the US context exist, albeit in a modified form, in the Nordic region.

Our discussion of metrosexuality in the media representations of six Nordic sportsmen highlights the concurrent existence of a *transnational metrosexual discourse* across national and regional boundaries, and a significant *international variation* that draws attention to *local specificity* and the fact that local contexts have not lost their significance. Our Nordic examples exhibit both similarities and differences to the US context as it appears in Coad's (2008) study. They also imply that metrosexuality, particularly its erotic aspect, is, in the Nordic region, to a somewhat lesser extent than in the United States, centered on non-Caucasian sportsmen.

NOTES

1. a) The proportion of material with focus on men in media coverage of sports varies between 78 and 92 percent in different studies (Ólafsson 2006; Pirinen 2006, 37–43).

   b) The concepts *Scandinavia* and the *Nordic countries* are in everyday talk often used interchangeably. Strictly speaking, however, the Nordic countries make up a wider entity including Iceland, Finland, Norway, Sweden, and Denmark, while Scandinavia is somewhat more limited covering only the latter three countries.

2. The rationale for selecting pairs that "represent" different countries is that each pair consists of two images that have been circulated in the same national field of publicity at roughly the same time. They have appeared in media that have been available for the same audiences and can thus be seen as representatives of discourses in the same public realm.

3. Sadly, we were unable to obtain permission to reproduce this image (and indeed, most of the images referred to in this chapter). All

web addresses referenced here were accessed successfully on June 7, 2013. This image of Haatinen can be found online at the following extremely long web address, or more simply by doing an Internet search with the terms "harri haatainen," "iltalehti," and "kuvagalleria." The photo first appeared in the Finnish newspaper *Iltalehti*.
http://www.google.com/imgres?imgurl=http://static.iltalehti.fi
/kuvagalleria/img/yleinen/6724.jpg&imgrefurl=http://www
.iltalehti.fi/kuvagalleria/data/yleinen/703/6.shtml&usg=__rXU
dUADtuoqO8dvTuzkwCZ7RoF4=&h=362&w=480&sz=53&hl
=en&start=6&zoom=1&tbnid=V77dFGPSxmEuGM:&tbnh=97
&tbnw=129&ei=5v6wUczxCqiO4ATVq4G4Cg&prev=/images
%3Fq%3D%2522harri%2Bhaatainen%2522%26client%3Dsafari%2
6sa%3DX%26rls%3Den%26hl%3Den%26tbm%3Disch&itbs=1&sa
=X&ved=0CDYQrQMwBQ.

4. This image was used in an advertising campaign for the sporting goods company Asics, and can be found as one of many images of Pöyhönen at http://www.queerclick.com/archive/2006/01 /queer_candy_mar.php, or with the logo cropped out, at http:// img57.imageshack.us/img57/1885/markusp07wq6.jpg.

5. The choice and analysis of the photographs have been informed by an earlier longitudinal study of masculinity constructions in representations of sportsmen in one of the two leading Finnish tabloids *Ilta-Sanomat* in 1983–2003 (see Wickman 2006, 151–156).

6. The quotes from Finnish, Swedish, and Norwegian media are translated into English by the authors.

7. See http://www.iltalehti.fi/kuvagalleria/data/yleinen/703/2.shtml.

8. The style of the drawing resembles the Sun Maid on packages of California raisins.

9. The picture was, however, repeatedly reprinted in the press as an illustration in appearance-centered articles on Haatainen or athletes more generally (*Iltalehti* May 17, 2003) gaining considerable circulation and visibility in the Finnish public in the following years.

10. In Norway, however, the women's national handball team, which has achieved more permanent success, has a stronger position in the national imaginary than the men's national handball team. But because of their progress and achievements in the mid-2000s, the most prominent players on the men's national handball team were, at that point in history, celebrated as national heroes.

11. The photograph was visible as a poster on the facades of more than a hundred tanning salons all around Norway. It was also used in several newspaper adverts.

12. Carew was the first soccer player on the Norwegian men's national team who was not completely white, but he has nevertheless often been portrayed as a "Norwegian" in the media. When he played for the Spanish team Valencia, however, he chose to use the Gambian middle name, Alieu, which aggravated many people (Andersson 2004, 1). He's constructed as both Norwegian and non-Norwegian at the same time.

13. The following analysis of the representation of Carew's body is a modified version of an analysis in an earlier work (see Langeland 2009, 51–53).

14. He caused at lot of attention by hitting a national team colleague, John Arne Riise, in 2004. In 2005, a paternity issue also figured in the Norwegian media. His controversial position was pointed out in the newspaper *Morgenbladet*, which claimed that Carew has been "persona non grata" in the media.

15. We were unable to obtain permission to reproduce the photographs from the *Dagbladet* interview that we analyze here. Unfortunately, they are also unavailable through the Internet, as the newspaper's online archives only date back to 2008. However, to get a sense of how John Carew's image in international media is hypersexualized, a simple Google search of his name, or a visit to the photo gallery on his personal website (http://www.jcarew .com/index.php/gallery) will suffice.

16. During the famous CK advertising campaign, a gigantic reproduction of this photo hung in a central spot in Sergels square in Stockholm, and it was deemed such a success that a revised version of it was made for a new campaign in 2005 in an alternative color scheme, black and red. The photo we analyze here can be found online at http://www.advertisingarchives.co.uk/index.php? service=search&action=do_quick_search&language=en&q =30532999. A similar photo from the same campaign appears here: http://metro.co.uk/2012/03/03/freddie-ljungberg-wanted -as-celebrity-big-brother-eye-candy-345674/.

17. In fact, this tattoo is reportedly a temporary reproduction of one of the two permanent ones that Ljungberg has on his back.

18. The photograph can be found online at the following sites: http://www.images99.com/sports/ soccer/zlatan-ibrahimovic -shirtless/ and http://habs.theoffside.com/gratuitous-nudity /zlatan-new-number-same-torso.html, among others.

REFERENCES

Ackerman, Sara L. 2010. "Plastic Paradise: Transforming Bodies and Selves in Costa Rica's Cosmetic Surgery Tourism Industry." *Medical Anthropology: Cross-Cultural Studies in Health and Illness* 29 (4): 403–23.

Adam, Barry D., Jan Willem Duyvendak, and André Krouwel. 1999. *The Global Emergence of Gay and Lesbian Politics: National Imprints of a Worldwide Movement.* Philadelphia, PA: Temple University Press.

Agathangelou, Anna M. 2004. *The Global Political Economy of Sex: Desire, Violence, and Insecurity in Mediterranean Nation States.* New York, NY: Palgrave Macmillan.

Agergaard, Sine. 2004. "Dansk kvindehåndbold i medierne: Fra 'jern-hårde ladies' til småpiger." *Dansk Sociologi* 15 (2): 91–106.

Akers Chacón, Justin. 2006. *No One Is Illegal: Fighting Violence and State Repression on the U.S.-Mexico Border.* Chicago, IL: Haymarket Books.

Akou, Heather Marie. 2004. "Nationalism without a Nation: Understanding the Dress of Somali Women in Minnesota." In *Fashioning Africa: Power and the Politics of Dress,* edited by Jean Allman, 50–63. Bloomington, IN: Indiana University Press.

Allison, Anne. 1994. *Nightwork: Sexuality, Pleasure, and Corporate Masculinity in a Tokyo Hostess Club.* Chicago: The University of Chicago Press.

Altman, Dennis. 1997. "Global Gaze/Global Gays." *GLQ: A Journal of Lesbian and Gay Studies* 3(3): 417–436.

Anderson, Benedict. 1983. *Imagined Communities: Reflections on the Origin and Spread of Nationalism.* London: Verso Random House.

Andersson, Mette. 2004. "Multikulturelle representanter mellom nasjonal og global toppidrett." idrottsforum.or: Nordic Sport Science Forum Oktober12. Accessed May 23, 2013. http://www.idrotts forum.org/articles/andersson/andersson041012.html.

Aparicio, Frances R. and Susana Chávez-Silverman, eds. 1997. *Tropicalizations: Transcultural Representations of Latinidad.* Hannover and London: Dartmouth University Press.

Arango Gaviria, Luz Gabriela. 2004. "Género, Trabajo e Identidad en los Estudios Latinoamericanos." In *Pensar (en) Género: Teoría y Práctica para Nuevas Cartografías del Cuerpo*, edited by Carmen Millán de Benavides and Angela María Estrada Mesa, 236–263. Bogotá, Colombia: Editorial Pontificia Universidad Javeriana.

Arnberg, Klara. 2009. "Synd på export: 1960-talets pornografiska press och den svenska synden." *Historisk tidskrift* 129 (3): 467–486.

Bach, Alice Riis. 2002. *Kvinder på banen—sport, køn og medier*. Copenhagen: Rosinante.

Balogun, Oluwakemi. 2012. "Cultural and Cosmopolitan: Idealized Femininity and Embodied Nationalism in Nigerian Beauty Pageants." *Gender & Society* 26 (3): 357–381.

Banner, Lois W. 1983. *American Beauty*. New York, NY: Knopf.

Basler, Carleen. R. 2012. "Latinas/os through the Lens." In *Cinematic Sociology: Social Life in Film*, edited by Jean-Anne Sutherland and Kathryn Feltey, 116–129. Thousand Oaks, CA: Sage.

Bautista López, Angélica and Elsa Conde Rodríguez. 2006. *Comercio sexual en La Merced: Una perspectiva constructivista sobre el sexoservicio*. Mexico City: Miguel Ángel Porrúa.

Bell, David, Ruth Holliday, Meredith Jones, Elspeth Probyn, and Jacqueline Sanchez Taylor. 2011. "Bikinis and Bandages: An Itinerary for Cosmetic Surgery Tourism." *Tourist Studies* 11 (2): 139–155.

Benton-Cohen, Katherine. 2009. *Borderline Americans: Racial Division and Labor War in the Arizona Borderlands*. London: Harvard University Press.

Bernstein, Elizabeth. 2007. *Temporarily Yours: Intimacy, Authenticity, and the Commerce of Sex*. Chicago, IL: University of Chicago Press.

Besteman, Catherine. 1995. "The Invention of Gosha: Slavery, Colonialism, and Stigma in Somali History." In *The Invention of Somalia*, edited by Ali J. Ahmed, 43–62. Lawrenceville, NJ: The Red Sea Press.

Bhabha, Homi. 1984. "Of Mimicry and Man: The Ambivalence of Colonial Discourse." *Discipleship: A Special Issue on Psychoanalysis* 28: 125–133.

Bird, Sharon R. 1996. "Welcome to the Men's Club: Homosociality and the Maintenance of Hegemonic Masculinity." *Gender & Society* 10: 120–132.

Black, Paula. 2004. *The Beauty Industry: Gender, Culture, Pleasure*. London: Routledge.

Bordo, Susan. 1999. *The Male Body*. New York, NY: Farrar, Straus & Giroux.

Boren, Michael Ian. 2006. "Important Places and Their Public Faces: Understanding Fenway Park as a Public Symbol." *The Journal of Popular Culture* 39: 205–224.

Bourdieu, Pierre. 1990. *The Logic of Practice.* Cambridge, MA: Polity.

Braun, Virginia. 2010. "Female Genital Cosmetic Surgery: A Critical Review of Current Knowledge and Contemporary Debates." *Journal of Women's Health* 19 (7): 1393–1407.

Brennan, Denise. 2004. *What's Love Got to Do with It? Transnational Desires and Sex Tourism in the Dominican Republic.* Durham, NC: Duke University Press.

Brents, Barbara G. and Kathryn Hausbeck. 2007. "Marketing Sex: U.S. Legal Brothels and Late Capitalist Consumption." *Sexualities* 10 (4):425–439.

Brewis, Joanna and Stephen Linstead. 1998. "Time after Time: The Temporal Organization of Red-Collar Work." *Time Society* 7 (2–3): 223–248.

Brewis, Joanna and Stephen Linstead. 2000. "'The Worst Thing Is the Screwing': Consumption and the Management of Identity in Sex Work." *Consumption and Identity* 7 (2): 84–97.

Bucciarelli, Fabio. 2010. "Nose Job: Iran, 2009," *Tehran Bureau: Frontline Newsmagazine,* May 26. Accessed August 1, 2012. http://www.pbs.org/wgbh/pages/frontline/tehranbureau/2010/nose-job.html.

Bureau of Labor Statistics. 2012. "Household Data Averages 11: Employed Persons by Detailed Occupation, Sex, Race, and Hispanic or Latino Ethnicity." Accessed June 3, 2013, http://www.bls.gov/cps/cpsaat11.pdf.

Cabezas, Amalia L. 2004. "Between Love and Money: Sex, Tourism, and Citizenship in Cuba and the Dominican Republic." *Signs: Journal of Women in Culture and Society* 29 (4): 987–1015.

Cabezas, Amalia L. 2009. *Economies of Desire: Sex and Tourism in Cuba and the Dominican Republic.* Philadelphia, PA: Temple University Press.

Candelario, Ginetta E. B. 2007. *Black behind the Ears: Dominican Racial Identity from Museums to Beauty Shops.* Durham, NC: Duke University Press.

Carrington, Ben. 2002. "Race, Representation and the Sporting Body." Critical Urban Studies: Occasional Papers. Centre for Urban and Community Research.

Casanova, Erynn Masi. 2004. "No Ugly Women": Concepts of Race and Beauty among Adolescent Women in Ecuador. *Gender & Society* 18 (3): 287–308.

Casanova, Erynn Masi. 2007. "The Whole Package: Exploring Cosmetic Surgery Tourism." Paper presented at the annual meeting of the American Sociological Association, New York, August 11.

Casanova, Erynn Masi. 2012. *Making up the Difference: Women, Beauty, and Direct Selling in Ecuador.* Austin, TX: University of Texas Press.

Castro-Gómez, Santiago and Eduardo Restrepo, eds. 2008. "Introducción: Colombianidad, población y diferencia." In *Genealogías de la Colombianidad: Formaciones Discursivas y Tecnologías de Gobierno en los Siglos XIX y XX.* Bogotá: Editorial Pontificia Universidad Javeriana, Instituto de Estudios Sociales y Culturales Pensar.

Chapkis, Wendy. 1986. *Beauty Secrets: Women and the Politics of Appearance.* Boston, MA: South End Press.

Choo, Hae Yeon. 2006. "Gendered Modernity and Ethnicized Citizenship: North Korean Settlers in Contemporary Korea." *Gender & Society* 20 (5): 576–604.

Classen, Constance, David Howes, and Anthony Synnott. 1994. *Aroma: The Cultural History of Smell.* London; New York, NY: Routledge.

Coad, David. 2008. *The Metrosexual: Gender, Sexuality, and Sport.* Albany, NY: State University of New York Press.

Cobble, Dorothy Sue. 1991. *Dishing It Out.* Chicago, IL: University of Illinois Press.

Cohen, Collen Ballerino, Richard Wilk, and Beverly Stoeltje. 1995. *Beauty Queens on the Global Stage: Gender, Contests, and Power.* New York, NY: Routledge.

Connell, R.W. 1995. *Masculinities.* Cambridge, MA: Polity Press.

Connell, Raewyn. 2009. *Gender: Short Introductions.* London: Polity.

Craig, Maxine Leeds. 2002. *Ain't I a Beauty Queen? Black Women, Beauty, and the Politics of Race.* New York, NY: Oxford University Press.

Cvajner, Martina. 2011. "Hyper-Femininity as Decency: Beauty, Womanhood and Respect in Emigration." *Ethnography* 12 (3): 356–374.

Dalstroma, Matthew D. 2012. "Winter Texans and the Re-creation of the American Medical Experience in Mexico." *Medical Anthropology: Cross-Cultural Studies in Health and Illness* 31 (2): 162–177.

Departamento Administrativo Nacional de Estadística. 2007. "Colombia Una Nación Multicultural: Su Diversidad Étnica." Dirección de Censos y Demografía. Accessed June 1, 2012. http://www.dane.gov.co/censo/files/presentaciones/grupos_etnicos.pdf.

Dewey, Susan. 2008. *Making Miss India Miss World: Constructing Gender, Power, and Nation in Postliberalization India*. New York, NY: Syracuse University Press.

Ditmore, Melissa. 2007. "In Calcutta, Sex Workers Organize." In *The Affective Turn: Theorizing the Social*, edited by Patricia Ticineto Clough, 170–186. Durham, NC: Duke University Press.

Dunn, Timothy J. 2009. *Blockading the Border and Human Rights: The El Paso Operation That Remade Immigration Enforcement*. Austin, TX: University of Texas Press.

Dyer, Richard. 1992. *Only Entertainment*. London: Routledge.

Dyer, Richard. 1997. *White*. London: Routledge.

Eco, Umberto. 2004. *The History of Beauty*. New York, NY: Rizzoli.

Edelman, Marc and Angelique Haugerud. 2004. "Introduction." In *The Anthropology of Development and Globalization: From Classical Political Economy to Contemporary Neoliberalism*, edited by Marc Edelman and Angelique Haugerud, 1–73. Malden, MA: Blackwell Publishing.

Edmonds, Alexander. 2010. *Pretty Modern: Beauty, Sex, and Plastic Surgery in Brazil*. Durham, NC: Duke University Press.

Edmonds, Alexander. 2011. "Almost Invisible Scars: Medical Tourism to Brazil." *Signs* 36 (2): 289–292.

Edwards, Tim. 1997. *Men in the Mirror: Men's Fashion, Masculinity and Consumer Society*. London: Cassell.

Enloe, Cynthia H. 2000. *Bananas, Beaches and Bases: Making Feminist Sense of International Politics*. Berkeley, CA: University of California Press.

Erickson, Karla. 2004. "Bodies at Work: Performing Service in American Restaurants." *Space and Culture* 7: 76–89.

Essen, Birgitta, and Sara Johnsdotter. 2004. "Female Genital Mutilation in the West: Traditional Circumcision versus Genital Cosmetic Surgery." *Acta Obstet Gynecol Scand* 83: 611–613.

Fernandes, Deepa. 2007. *Targeted: Homeland Security and the Business of Immigration*. London: Seven Stories.

Flood, Michael. 2008. "Men, Sex, and Homosociality: How Bonds between Men Shape Their Sexual Relations with Women." *Men and Masculinities* 10: 339–359.

Fojas, Camilla. 2008. "New Frontiers of Asian and Latino America in Popular Culture: Mixed-Race Intimacies and the Global Police State in Miami Vice and Rush Hour 2." *Journal of Asian American Studies* 14 (3): 417–434.

Friedman, Jonathan. 2004. "Globalization, Dis-integration, Re-organization: The Transformations of Violence." In *The*

Anthropology of Development and Globalization: From Classical Political Economy to Contemporary Neoliberalism, edited by Marc Edelman and Angelique Haugerud, 160–168. Malden, MA: Blackwell Publishing.

Gal, Susan and Gail Kligman. 2000. *The Politics of Gender after Socialism: A Comparative-Historical Essay*. Princeton, NJ: Princeton University Press.

Gellert, Tamas. 2010. "För Sverigedemokraterna är Zlatan inte svensk." *Dagens Nyheter: Kultur & Nöje* September 17. Accessed August 8, 2012. http://www.dn.se/kultur-noje/debatt-essa/for-sverigedemokraterna-ar-zlatan-inte-svensk-1.1172192.

Gill, Rosalind 2009. "Beyond the 'Sexualization of Culture' Thesis: An Intersectional Analysis of 'Sixpacks,' 'Midriffs' and 'Hot Lesbians' in Advertising." *Sexualities* 12 (2):137–160.

Gilman, Sander L. 1999. *Making the Body Beautiful: A Cultural History of Aesthetic Surgery*. Princeton, NJ: Princeton University Press.

Gimlin, Debra L. 2002. *Body Work: Beauty and Self-Image in American Culture*. Berkeley, CA: University of California Press.

Gimlin, Debra. 2007. "What Is 'Body Work'? A Review of the Literature." *Sociological Compass* 1: 353–370.

Glenn, Evelyn Nakano, ed. 2009. *Shades of Difference: Why Skin Color Matters*. Stanford, CA: Stanford University Press.

González Alafita, María Eugenia; Cristina Dávalos, and Mariell Gutiérrez. 2012. "Modern Family y los Mensajes Culturales: Percepciones de Jóvenes Receptores Mexicanos de la Serie Televisiva Estadounidense." *Revista Comunicación* 10: 517–530.

Gordon, Tuula. 2002. "Kansalisuuden fyysiset, sosiaaliset ja mentaaliset rajat." In (ed.) *Sukupuolitetut rajat: Gendered Borders and Boundaries (Psykologian tutkimuksia 22)*, edited by Katri Komulainen, 15–37. Joensuu: University of Joensuu.

Gruenbaum, Ellen. 2001. *The Female Circumcision Controversy: An Anthropological Perspective*. Philadelphia, PA: University of Pennsylvania Press.

GSO Vietnam. 2011. *Foreign Direct Investment Projects Licensed in Period 1988–2010*, edited by Investment Department. Hanoi: General Statistics Office if Vietnam.

Gullestad, Marianne. 2007. *Misjonsbilder. Bidrag til norsk selvforståelse: om bruk av foto og film i tverrkulturell kommunikasjon*. Oslo: Universitetsforlaget.

Guzmán, Isabel Molina and Angharad N. Valdivia. 2004. "Brain, Brow, and Booty: Latina Iconicity in U.S. Popular Culture." *The Communication Review* 7: 205–221.

Hall, Stuart and Paul du Gay, eds. 1996. *Questions of Cultural Identity*. London: Sage Publications.

Hamermesh, Daniel S. 2011. *Beauty Pays*. Princeton, NJ: Princeton University Press.

Hausmann, Ricardo, Laura D. Tyson, and Saadia Zahidi. 2010. *The Global Gender Gap Report 2010*. Geneva, SU: The World Economic Forum.

Hayton, Bill. 2010. *Vietnam: Rising Dragon*. New Haven, CT: Yale University Press.

Hearn, H. L. and Patricia Stoll. 1975. "Continuance Commitment in Low-Status Occupations: The Cocktail Waitress." *The Sociological Quarterly* 16: 105–114.

Hedetoft, Ulf. 2003. *The Global Turn: National Encounters with the World*. Aalborg, DK: Aalborg University Press.

Hemmingson, Michael. 2008. *Zona Norte: The Post-Structural Body of Erotic Dancers and Sex Workers in Tijuana, San Diego and Los Angeles: An Auto/Ethnography of Desire and Addiction*. Newcastle: Cambridge Scholars.

Hernlund, Ylva and Bettina Shell-Duncan. 2007. "Transcultural Positions: Negotiating Rights and Culture." In *Transcultural Bodies: Female Genital Cutting in Global Context*, edited by Ylva Hernlund and Bettina Shell-Duncan, 1–45. New Brunswick, NJ; London: Rutgers University Press.

Herzog, Lawrence A. 2003. "Global Tijuana." In *Postborder City: Cultural Spaces of Bajalta California*, edited by Gustavo Leclerc and Michael Dear, 119–142. New York, NY: Routledge.

Hesse-Biber, Sharlene. 1997. *Am I Thin Enough Yet? The Cult of Thinness and the Commercialization of Identity*. New York, NY: Oxford University Press.

Hietala, Veijo. 2003. "Painii myyttien kanssa—Urheilu modernissa ja postmodernissa mediakulttuurissa." *Lähikuva* 1: 6–15.

Hill, Mark. 2000. "Color Differences in the Socioeconomic Status of African American Men: Results of a Longitudinal Study." *Social Forces* 78 (4): 1437–1460.

Hill, Mark. 2002. "Skin Color and the Perception of Attractiveness among African Americans: Does Gender Make a Difference?" *Social Psychology Quarterly* 65 (1): 77–91.

Hill Collins, Patricia. 2004. *Black Sexual Politics: African Americans, Gender, and the New Racism*. New York, NY: Routledge.

Hoang, Kimberly. 2011. "New Economies of Sex and Intimacy in Vietnam." PhD diss., University of California—Berkeley.

Hochschild, Arlie. 1983. *The Managed Heart: Commercialization of Human Feeling*. Berkeley, CA: University of California Press.

Hofmann, Susanne. 2005. "Life Histories of Sex Workers in Mexico City: An Ethnological Study on Experiences of Gender Violence." MA diss., Free University of Berlin.

Hofmann, Susanne. 2010. "Corporeal Entrepreneurialism and Neoliberal Agency in the Sex Trade at the US-Mexican Border." *Women's Studies Quarterly* 38 (3/4): 233–256.

Holguin, Jaime. 2007. "Iran: Nose Job Capital of the World," in *CBS Evening News*, February 11. Accessed June 1, 2013. http://www.cbsnews.com/8301–18563_162–692495.html.

Holliday, Ruth and Joanna Elfving-Hwang. 2012. "Gender, Globalization and Aesthetic Surgery in South Korea." *Body and Society* 18 (2): 58–81.

Hooters of America. 2007. "Hooters Facts." Hooters of America website: www.hooters.com/Company/Didyouknow.aspx. Accessed June 28, 2009.

Huisman, Kimberly and Pierrette Hondagneu-Sotelo. 2005. "Dress Matters: Change and Continuity in the Dress Practices of Bosnian Muslim Refugee Women." *Gender & Society* 19 (1): 44–65.

Human Rights Watch. 2010. "Welcome to Kenya": Police Abuse of Somali Refugees. Accessed March 16, 2012. http://www.hrw.org/sites/default/files/reports/kenya0610webwcover.pdf.

Hunter, Margaret L. 1998. "Colorstruck: Skin Color Stratification in the Lives of African American Women." *Sociological Inquiry* 68 (4): 517–535.

Hunter, Margaret. 2002. "If You're Light, You're Alright: Light Skin Color as Social Capital for Women of Color." *Gender & Society* 16 (2): 175–193.

Hunter, Margaret. 2005. *Race, Gender and the Politics of Skin Tone.* New York, NY: Routledge.

Inda, Jonathan X. and Renato Rosaldo. 2008. "Tracking Global Flows." In *The Anthropology of Globalization: A Reader* (Second Edition), edited by Jonathan Xavier Inda and Renato Rosaldo, 3–46. Malden, MA; Oxford: Blackwell Publishing.

International Society of Aesthetic Plastic Surgeons (ISAPS). "International Survey on Aesthetic/Cosmetic Procedures Performed in 2011." Accessed February 20, 2013. http://www.isaps.org/files/html-contents/Downloads/ISAPS%20Results%20-%20Procedures%20in%202011.pdf.

Isotalo, Anu. 2007. ""Did You See Her Standing at the Marketplace?" Gender, Gossip, and Socio-Spatial behaviour of Somali girls in Turku, Finland." In *From Mogadishu to Dixon: the Somali Diaspora in a Global Context*, edited by Abdi M. Kusow, 181–206. Trenton, NJ; Asmara, Eritrea: The Red Sea Press.

Jafar, Afshan. 2012. "Progress and Women's Bodies." TEDxTalks, 23:36. Posted May 7. http://www.youtube.com/watch?v=Baxnv wffWbE.

James, Stanlie M. and Claire C. Robertson. 2002. *Genital Cutting and Transnational Sisterhood: Disputing U.S. Polemics.* Urbana; Chicago, IL: University of Illinois Press.

Jaramillo, Patricia and Vivian Nayibe Castro Romero. 2009. "Comercio, Mercado Laboral y Economía del Cuidado en Colombia: Propuesta de política pública." *La Red Internacional de Género y Comercio.* Accessed February 20, 2013. http://www.gen eroycomercio.org/areas/incidencia/Policy_Paper_Colombia.pdf.

Jeffreys, Sheila. 2005. *Beauty and Misogyny: Harmful Cultural Practices in the West.* London: Routledge.

Jeffords, Susan. 1994. *Hard Bodies: Hollywood Masculinity in the Reagan Era.* New Brunswick, NJ: Rutgers University Press.

Jimeno, Myriam, Andrés Góngora, Marco Martínez, and Carlos José Suárez, eds. 2007. *Manes, Mancitos y Manazos: Una metodología de trabajo sobre violencia intrafamiliar y sexual. Colección CES: Serie Conflicto, Violencia y Sociedad.* Bogotá, Colombia: Facultad de Ciencias Humanas, Centro de Estudios Sociales, Universidad Nacional de Colombia.

Johansson, Thomas. 1998. "Muskler, svett och maskulinitet." In *Rädd att falla: Studier i manlighet*, edited by Ekenstam, Clas, Jonas Fryckman, and Thomas Johansson, 244–268. Hedemora /Möklinta: Gidlunds förlag.

Jones, Geoffrey. 2010. *Beauty Imagined: A History of the Global Beauty Industry.* New York, NY: Oxford University Press.

Jones, Geoffrey. 2011. "Globalization and Beauty: A Historical and Firm Perspective." *EurAmerica* 41 (4): 885–916.

Kanaaneh, Rhoda. A. 2002. *Birthing the Nation: Strategies of Palestinian Women in Israel.* Berkeley; Los Angeles; London: University of California Press.

Karim, Persis M. 2010. "In Praise of Big Noses." *The Atlanta Review* 16 (2): 45–46.

Kempadoo, Kamala. 1999. *Sun, Sex, and Gold: Tourism and Sex Work in the Caribbean.* Lanham, MD: Rowman and Littlefield Publishers.

Kiesling, Scott Fabious. 2005. "Homosocial Desire in Men's Talk: Balancing and Re-Creating Cultural Discourses of Masculinity." *Language in Society* 34: 695–726.

King-O'Riain, Rebecca Chiyoko. 2006. *Pure Beauty: Judging Race in Japanese American Beauty Pageants*. Minneapolis, MN: University of Minnesota Press.

King-O'Riain, Rebecca Chiyoko. 2008. "Making the Perfect Queen: The Cultural Production of Identities in Beauty Pageants." *Sociology Compass* 2 (1): 74–83.

Kinnunen, Taina. 2003. "Kansakunnan sotureita ja ihannevartaloita—television urheilu-uutisten miesruumiin representaatiot." *Lähikuva* 1: 16–29.

Kinnunen, Taina and Jan Wickman. 2007. "Pin-Up Warriors." *NORMA: The Nordic Journal of Masculinity Studies* 2 (1): 167–181.

Kjeldstad, Randi. 2001. "Gender Policies and Gender Equality." In *Nordic Welfare States in the European Context*, edited by Kautto, Mikko, Johan Fritzell, Björn Hvinden, Jon Kvist, and Hannu Uusitalo, 66–97. London: Routledge.

Koskela, Hille. 2010. "Did You Spot an Alien? Voluntary Vigilance, Borderwork and the Texas Virtual Border Watch Program." *Space & Polity* 14 (2): 103–121.

Kuczynski, Alex. 2007. *Beauty Junkies: Under the Skin of the Cosmetic Surgery Industry*. London: Vermilion.

Landes, Joan B. 2001. *Visualizing the Nation: Gender, Representation and the Revolution in Eighteenth-Century France*. Ithaca, NY: Cornell University Press.

Langeland, Fredrik. 2009. "Den norske kroppen." In *Norske seksualiteter*, edited by Wencke Mühleisen, and Åse Røthing, 37–58. Oslo: Cappelen.

Langeland, Fredrik. 2011. "Maskulinitetens refleksive nostalgi i Tv2 Zebras Manshow." *Tidsskrift for kjønnsforskning* 35(4): 275–292.

Letmark, Peter. 2011. "Mannen gör sig till sex objekt." *Dagens Nyheter* June 9: 22–23.

Lever, Janet and Deanne Dolnick. 2010. "Call Girls and Street Prostitutes: Selling Sex and Intimacy." In *Sex for Sale: Prostitution, Pornography, and the Sex Industry*, edited by Ronald Weitzer, 187–204. New York, NY: Routledge.

Liao, Lih-Mei, Neda Taghinejadi, and Sarah M. Creighton. 2012. "An Analysis of the Content and Clinical Implications of Online Advertisements for Female Genital Cosmetic Surgery." *Obstetrics and Gynaecology* 2. Accessed June 6, 2013. doi:10.1136/bmjopen-2012–001908.

Li, Eric, Hyun Min, Russell Belk, Junku Kimura, Shalini Bahl. 2008. "Skin Lightening and Beauty in Four Asian Cultures." *Advances in Consumer Research* 35: 444–449.

Lindley, Anna. 2011. "Between a Protracted and Crisis Situation: Policy Responses to Somali Refugees in Kenya." *Refugee Survey Quarterly* 30 (4): 14–49.

Lippe, Gerd, von der. 2002. "Media Image: Sport, Gender and National Identities in Five European Countries." *International Review for the Sociology of Sport* 37 (3/4): 371–395.

Lippe, Gerd von der. 2010. "Et medieskapt maskulinit begjærs—og idrettsidol." In *Retorikk, idrett og samfunn*, edited by Hans-Ivar Kristiansen and Odd Nordhaug, 155–180. Oslo: Forlag1.

Loe, Meika. 1996. "Working for Men at the Intersection of Power, Gender, and Sexuality." *Sociological Inquiry* 66 (4): 399–422.

Low, Kelvin. 2006. "Presenting the Self, the Social Body, and the Olfactory: Managing the Self in Everyday Life Experiences." *Sociological Perspectives* 49 (4): 607–631.

Manalansan, Martin. 2006. "Immigrant Lives and the Politics of Olfaction in the Global City." In *The Smell Culture Reader*, edited by Jim Drobnick, 41–52. Oxford and New York, NY: Berg.

Mani, Lata. 1998. *Contentious Traditions: The Debate on Sati in Colonial India*. Berkeley, CA: University of California Press.

Menchaca, Martha. 2001. *Recovering History, Constructing Race: The Indian, Black and White Roots of Mexican Americans*. Austin, TX: University of Texas Press.

Messner, Michael and Donald Sabo. 1990. *Sport, Men and the Gender Order: Critical Feminist Perspectives*. Champaign, IL: Human Kinetics.

Mears, Ashley. 2010. "Size Zero High-End Ethnic: Cultural Production and the Reproduction of Culture in Fashion Modeling." *Poetics* 38 (1): 21–46.

Moeran, Brian. 2007. "Marketing Scents and the Anthropology of Smell." *Social Anthropology* 15 (2): 153–168.

Mühleisen, Wencke and Åse Røthing, eds. 2009. *Norske seksualiteter*. Oslo: Cappelen.

Mulinari, Diana. 2008. "Women Friendly? Understanding Gendered Racism in Sweden." In *Gender Equality and Welfare Politics in Scandinavia: The Limits of Political Ambition?* edited by Kari Melby, Anna-Birte Ravn, and Christina Carlsson Wetterberg, 167–182. Bristol: Policy Press.

Mulvey, Laura. 1975. "Visual Pleasure and Narrative Cinema." *Screen* 16 (3): 6–18.

Muriá Tuñón, Magalí. 2010. "Enforcing Boundaries: Globalization, State Power and the Geography of Cross-Border Consumption in Tijuana, Mexico." PhD diss., University of California, San Diego.

Negrón-Muntaner, Frances. 2000. "Feeling Pretty: West Side Story and Puerto Rican identity Discourses." *Social Text* 63: 83–106.

Negrón-Muntaner, Frances. 2004. *Boricua Pop: Puerto Ricans and the Latinization of American Culture*. New York, NY: New York University Press.

Nevins, Joseph. 2002. Operation Gatekeeper: *The Rise of the Illegal Alien and the Making of the U.S.-Mexico Boundary*. London: Routledge.

Newton-Francis, Michelle. 2008. "The Hooters Girl and the Conundrum of Connotations: An Exploratory Study of the Use of the Cultural Toolkit in Managing Stigma." PhD. diss., American University, Washington, DC, 2008.

Nnaemeka, Obioma. 2005. *Female Circumcision and the Politics of Knowledge: African Women in Imperialist Discourses*. Westport, CT; London: Praeger.

O'Brien, Lucy. 2002. She Bop: *The Definitive History of Women in Rock, Pop, and Soul*. New York, NY: Continuum.

Ochoa, Marcia. 2005. "Queen for a Day: Transformistas, Misses and Mass Media in Venezuela." PhD diss., Stanford University.

O'Connell Davidson, Julia and Jaqueline Sanchez Taylor. 1999. "Fantasy Islands: Exploring the Demand for Sex Tourism." In *Sun, Sex, and Gold: Tourism and Sex Work in the Caribbean*, edited by Kamala Kempadoo, 37–54. Lanham, MD: Rowman and Littlefield Publishers.

O'Connell Davidson, Julia. 1998. *Prostitution, Power and Freedom*. London: Polity.

Ólafsson, Kjartan, ed. 2006. *Sport, Media and Stereotypes: Women and Men in Sports and Media*. Akureyri: Centre for Gender Equality. http://www.gender.is/sms.

O'Neill, Maggie. 1996. "The Aestheticization of the Whore in Contemporary Society: Desire, the Body, Self and Society." Paper presented to the Body and Organization Workshop, Keele University, Keele, September.

Ong, Aihwa. 2008. "Cyberpublics and Diaspora Politics among Transnational Chinese." *In The Anthropology of Globalization: A Reader* (Second Edition), edited by Jonathan Xavier Inda and Renato Rosaldo, 167–183. Malden, MA; Oxford: Blackwell Publishing.

Ossman, Susan. 2002. *Three Faces of Beauty: Casablanca, Paris, Cairo*. Durham, NC: Duke University Press.

Otis, Eileen. 2012. *Markets and Bodies: Women, Service Work, and the Making of Inequality in China*. Stanford, CA: Stanford University Press.

O'Toole, Brockton. 2013. "Brockton O'Toole Tijuana FAQ." Accessed May 4. http://members.clubhombre.com/brockton/.

Palacio Valencia, María Cristina, and Ana Judith Valencia Hoyos. 2001. *La identidad masculina: Un mundo de Inclusiones y Exclusiones. Colección Ciencias Jurídicas y Sociales.* Manizales, Colombia: Editorial Universidad de Caldas,.

Peiss, Kathy. 1998. *Hope in a Jar: The Making of America's Beauty Culture.* New York, NY: Metropolitan Books.

Pierre, Jemima. 2008. "'I Like Your Colour!' Skin Bleaching and Geographies of Race in Urban Ghana." *Feminist Review* 90 (1): 9–29.

Pinho, Patricia de Santana. 2009. "White but Not Quite: Tones and Overtones of Whiteness in Brazil." *Small Axe* 29: 40–56.

Pirinen, Riitta. 2006. *Urheileva Nainen lehtiteksteissä.* Tampere: Tampere University Press. Online: Acta Electronica Universitatis Tamperensis 512. Accessed December 13, 2012. http://acta.uta.fi/pdf/951-44-6574-1.pdf.

Plummer, Ken. 1998. Preface to *Scandinavian Homosexualities: Essays on Gay and Lesbian Studies,* edited by Jan Löfström, xiii–xvi. Binghamton: Harrington Park Press.

Poole, Deborah. 2011. "Mestizaje, Distinction, and Cultural Presence: The View from Oaxaca." In *Histories of Race and Racism: The Andes and Mesoamerica from Colonial Times to the Present,* edited by Laura Gotkowitz, 179–203. Durham, NC: Duke University Press.

Rahier, Jean Muteba. 1998. "Blackness, the Racial/Spatial Order, Migrations, and Miss Ecuador 1995–96." *American Anthropologist* 100 (2): 421–430.

Rahier, Jean Muteba. 1999. "Body Politics in Black and White: Señoras, Mujeres, Blanqueamiento and Miss Esmeraldas 1997–1998, Ecuador." *Women and Performance: A Journal of Feminist Theory* 11 (1): 103–120.

Rahman, Anika and Nahid Toubia. 2000. *Female Genital Mutilation: A Guide to Laws and Policies Worldwide.* London: Zed Books.

Rambo Ronai, Carol and Carolyn Ellis. 1989. "Turn Ons for Money: Interactional Strategies of the Table Dancer." *Journal of Contemporary Ethnography* 18 (3): 271–298.

Ramírez de Arellano, Annette B. 2011. "Medical Tourism in the Caribbean." *Signs* 36 (2): 289–297.

Rasmussen, Susan. 1999. "Making Better 'Scents' in Anthropology: Aroma in Tuareg Sociocultural Systems and the Shaping of Ethnography." *Anthropological Quarterly* 72 (2): 55–73.

Rasmusson, Sarah L. 2011. "'We're Real Here:' Hooters Girls, Big Tips, & Provocative Research Methods." *Cultural Studies <-> Critical Methodologies*, 11: 574–585.

Reitman, Meredith. 2006. "Uncovering the White Place: White Washing at Work." *Social and Cultural Geography* 7 (2): 267–282.

Restaurants and Institutions. "R&I 2009 Top 400 Restaurants Chains." *Restaurants and Institutions Trade Magazine*, July 15, 2009.

Rhode, Deborah L. 2010. *The Beauty Bias: The Injustice of Appearance in Life and Law.* New York, NY: Oxford University Press.

Rofel, Lisa. 1999. *Other Modernities: Gendered Yearnings in China after Socialism.* Berkeley, CA: University of California Press.

Rogers, Mark. 1998. "Spectacular Bodies: Folkorization and the Politics of Identity in Ecuadorian Beauty Pageants." *Journal of Latin American Anthropology* 3 (2): 54–85.

Rohlinger, Deana A. 2002. "Eroticizing Men: Cultural Influences on Advertising and Male Objectification." *Sex Roles: A Journal of Research* 46 (2): 61–74.

Romero, Fernando. 2008. *Hyperborder: The Contemporary U.S.-Mexico Border and Its Future.* New York, NY: Princeton Architectural Press.

Røthing, Åse, and Stine Helena Svendsen. 2009. "Norskhet og seksualitet i skolen." In *Norske seksualiteter*, edited by Wencke Mühleisen, and Åse Røthing, 59–78. Oslo: Cappelen.

Ruiz Marín, Elvia Lucía, Claudia Emilse López Aristizábal, and Juan Gonzalo Escobar Correa 2011. "Los Jóvenes, el ideal estético, y la televisión: El cuerpo real y el imaginado." *Revista Luciérnaga* 3:17–22.

Russell, Kathy, Midge Wilson, and Ronald Hall. 1992. *The Color Complex: The Politics of Skin Color Among African Americans.* New York, NY: Harcourt Brace Jovanovich.

Ryan, Chris and Colin Michael Hall. 2001. *Sex Tourism: Marginal People and Liminalities.* London: Routledge.

Said, Edward. 1978. *Orientalism.* New York: Pantheon.

Sanders, Teela. 2005a. "'It's Just Acting': Sex Workers' Strategies for Capitalizing on Sexuality." *Gender, Work and Organization* 12 (4): 319–342.

Sanders, Teela. 2005b. *Sex Work: A Risky Business.* Devon: Willan Publishing.

Sanders, Teela. 2008. "Male Sexual Scripts: Intimacy, Sexuality and Pleasure in the Purchase of Commercial Sex." *Sociology* 42: 400–419.

Saraswati, L. Ayu. 2010. "Cosmopolitan Whiteness: The Effects of and Affects of Skin-Whitening Advertisements in a Transnational

Women's Magazine in Indonesia." *Meridians: Feminism, Race, Transnationalism* 10 (2): 15–41.

Schlack, Jennifer Daryl. 1996. "The Theory and Method of Articulation in Cultural Studies." In *Stuart Hall: Critical Dialogues in Cultural Studies*, edited by David Morley and Chen Huan-Hsing, 113–130. London: Routledge.

Shell-Duncan, Bettina, and Ylva Hernlund. 2000. *Female Circumcision in Africa: Culture, Controversy, and Change*. Boulder, CO; London: Lynne Rienner Publishers.

Shilling, Chris. 2003. *The Body and Social Theory*, 2nd edition. Thousand Oaks, CA: Sage Publications.

Simpson, Mark. 1994. "Here Come the Mirror Men." *The Independent*. November 15. Accessed August 15, 2012. http://www.marksimp son.com/pages/journalism/mirror_men.html.

Simpson, Mark. 2002. "Meet the Metrosexual." Salon: Arts & Entertainment July 22. Accessed May 24, 2013. http://www.salon .com/entfeature/2002/07/22/metrosexual.

Singer, Linda. 1993. *Erotic Welfare: Sexual Theory and Politics in the Age of Epidemic*. London: Routledge.

Skille, Eivind Å. 2006. "Forskning om kjønn, etnisitet og ungdom som økt forståelse for forskjeller i ungdoms idrettsdeltakelse." idrottsforum.org: Nordic Sport Science Forum September 13. Acessed August 15, 2012. http://idrottsforum.org/articles/skille /skille060913.pdf.

Sklair, Leslie. 2001. *The Transnational Capitalist Class*. Oxford: Blackwell.

Soto, Rosa E. 2008. "Looking Latina: Cultural Perspectives on Images and Representations of Latinas in Film, Television and Popular Culture." PhD diss., University of Florida.

Strandbu, Åse. 2005. "Identity, Embodied Culture and Physical Exercise: Stories from Muslim Girls in Oslo with Immigrant Backgrounds." *Young: Nordic Journal of Youth Research* 13 (1): 27–45.

Sullivan, Nikki. 2007. "'The Price to Pay for Our Common Good': Genital Modification and the Somatechnologies of Cultural (In) Difference." *Social Semiotics* 17 (3): 395–409.

Talle, Aud. 1993. "Transforming Women into 'Pure' Agnates: Aspects of Female Infibulation in Somalia." In *Carved Flesh/ Cast Selves: Gendered Symbols and Social Practices*, edited by Vigdis Broch-Due, Ingrid Rudie, and Tone Bleie, 83–106. Oxford: Berg Publishers.

Talle, Aud. 2007. "Female Circumcision in Africa and Beyond: The Anthropology of a Difficult Issue." In *Transcultural Bodies: Female Genital Cutting in Global Context*, edited by Ylva Hernlund and

Bettina Shell-Duncan, 91–106. New Brunswick, NJ; London: Rutgers University Press.

Talukdar, Jaita. 2012. "Thin but Not 'Skinny': Women Negotiating the 'Never Too Thin' Body Ideal in Urban India." *Women's Studies International Forum* 35: 109–118.

Talukdar, Jaita and Annulla Linders. 2013. "Gender, Class Aspirations, and Emerging Fields of Body Work in Urban India." *Qualitative Sociology* 36: 101–123.

Tate, Shirley A. 2009. *Black Beauty: Aesthetics, Stylization, Politics.* Farnham: Ashgate.

Taussig, Michael. 2008. "La Bella y la Bestia/Beauty and the Beast." (Translated by Sally Station.) *Antípoda*, 6: 17–40.

Tervo, Mervi. 2002. "Sports, 'Race' and the Finnish National Identity in Helsingin Sanomat in the Early Twentieth Century." *Nations and Nationalism* 8 (3): 335–356.

Thompson, Maxine S. and Verna Keith. 2001. "The Blacker the Berry: Gender, Skin Tone, Self-Esteem, and Self-Efficacy." *Gender & Society* 15: 336–357.

Tibbals, Chauntelle Anne. 2007. "Doing Gender as Resistance: Waitresses and Servers in Contemporary Table Service." *Journal of Contemporary Ethnography* 36: 731–751.

Tiilikainen, Marja. 2007. "Continuity and Change: Somali Women and Everyday Islam in the Diaspora." In *From Mogadishu to Dixon: The Somali Diaspora in a Global Context*, edited by Abdi M. Kusow, 207–232. Trenton, NJ; Asmara, Eritrea: The Red Sea Press.

Tuori, Salla, Suvi Keskinen, Sari Irni, and Diana Mulinari, eds. 2009. *Complying with Colonialism: Gender, Race and Ethnicity in the Nordic Region.* Farnham: Ashgate.

Turner, Bryan. 1996. *The Body and Society: Explorations in Social Theory.* London: Sage Publications.

Tyler, Melissa. 2012. "Working in the Other Square Mile: Performing and Placing Sexualized Labour." *Work Employment Society* 26: 899–916.

Vaid, Jyotsna. 2009. "Fair Enough? Color and the Commodification of Self in Indian Matrimonials." In *Shades of Difference: Why Skin Color Matters*, edited by Evelyn Nakano Glenn, 148–165. Stanford, CA: Stanford University Press.

Valdivia, Angharad N. 1998. "Stereotype or Transgression? Rosie Perez in Hollywood Film." *The Sociological Quarterly* 39: 393–408.

Viveros Vigoya, Mara. 2002. *De Quebradores y Cumplidores: Sobre Hombres, Masculinidades y Relaciones de Género en Colombia.* Centro de Estudios Sociales: Universidad Nacional.

vom Bruck, Gabriele. 2002. "Elusive Bodies: The Politics of Aesthetics among Yemeni Elite Women." In *Gender, Politics, and Islam*, edited by Therese Saliba, Carolyn Allen, and Judith A. Howard, 161–200. Chicago; London: The University of Chicago Press.

Wacquant, Loïc. 1995. "Pugs at Work: Bodily Capital and Bodily Labour among Professional Boxers." *Body & Society* 1 (1): 65–93.

Wacquant, Loïc. 2001. "Whores, Slaves and Stallions: Languages of Exploitation and Accommodation among Boxers." *Body & Society* 7(2–3): 181–194.

Wacquant, Loïc. 2004. *Body and Soul: Notebooks of an Apprentice Boxer*. Oxford: Oxford University Press.

Wagatsuma, Hiroshi. 1967. "The Social Perception of Skin Color in Japan." *Daedalus* 96 (2): 407–443.

Warren, Roger. 1990. *Red Lights of Baja Mexico: A Guide to Baja's Zones of Tolerance*. San Diego: Warren Communications.

Weinbaum, Alys Eve, Lynn M. Thomas, Priti Ramamurthy, Uta G. Poiger, Madeleine Y. Dong, and Tani E. Barlow. 2008. "The Modern Girl as a Heuristic Device: Collaboration, Connective Comparison, Multidirectional Citation." In *The Modern Girl around the World: Consumption, Modernity, and Globalization*, edited by Alys Eve Weinbaum, Lynn M. Thomas, Priti Ramamurthy, Uta G. Poiger, Madeleine Y. Dong, and Tani E. Barlow. Durham, NC: Duke University Press.

Wellington, Christine A. and John R. Bryson. 2001. "At Face Value? Image Consultancy, Emotional Labour and Professional Work." *Sociology* 35 (4): 933–946.

Whannel, Garry 2002. *Media Sport Stars: Masculinities and Moralities*. London: Routledge.

Wickman, Jan 2004. "Ortodox maskulinitet versus metrosexualitet—spänningar i finländska mediarepresentationer av idrottsmän." Paper presented at the Nordic Conference on Men and Masculinities: Den gode, den onde, den normale, Södertälje, Sweden, November 26–28. Accessed August 15, 2012. http://www.kvinfo.su.se/lankar/mansforskn%20konf%20-%20Wickman.pdf.

Wickman, Jan. 2006. "Mediaseksikäs miesurheilija." In *Seksuaalinen ruumis*, edited by Taina Kinnunen, and Anne Puuronen, 143–159. Helsinki, FI: Gaudeamus.

Wickman, Jan. 2011. "Review of *The Metrosexual: Gender, Sexuality and Sport* by David Coad." *Men and Masculinities* 14(1): 117–119.

Williamson, Judith. 1986. "Woman is an Island: Femininity and Colonization." In *Studies in Entertainment*, edited by Tania Modleski, 99–118. Bloomington, IN: Indiana University Press.

WHO. 2008. Eliminating Female Genital Mutilation: An Interagency Statement. Geneva; Switzerland: WHO Press. (available online at http://whqlibdoc.who.int/publications/2008/9789241596442 _eng.pdf).

Wonders, Nancy A. and Raymond Michalowski. 2001. "Bodies, Borders, and Sex Tourism in a Globalized World: A Tale of Two Cities—Amsterdam and Havana." *Social Problems* 48 (4): 545–571.

Yee, Daniel. 2006. "Waitresses Help Hooters Restaurants Fly." *The Associated Press and State Local Wire*. December, 14.

Zheng, Tiantian. 2009. *Red Lights: The Lives of Sex Workers in Postsocialist China*. Minneapolis: University of Minnesota Press.

Zoonen, Liesbet van. 1994. *Feminist Media Studies*. London: Sage Publications.

CONTRIBUTORS

Oluwakemi M. Balogun is an assistant professor of Sociology and Women's and Gender Studies at the University of Oregon and was formerly a postdoctoral fellow at Pomona College. She received her PhD in Sociology with a Designated Emphasis in Women, Gender, and Sexuality from the University of California Berkeley in 2012. She has published articles in outlets such as *Ethnicities* and *Gender & Society*. Her work focuses on gender, globalization, nationalism, ethnicity, and migration. She is currently working on a book that examines the Nigerian beauty pageant industry in order to document the country's transition from post-independence to an emerging nation.

Kaija Bergen is a Macalester College graduate with degrees in English and International Studies. Born in Colorado, she served as a US Peace Corps volunteer in Cambodia from 2011 to 2013, where she taught English at the Kampot Provincial Teacher Training College.

Erynn Masi de Casanova is an assistant professor of Sociology at the University of Cincinnati, where she is also a faculty affiliate of the Department of Women's, Gender, and Sexuality Studies and the Department of Romance Languages and Literatures. She is the author of *Making Up the Difference: Women, Beauty, and Direct Selling in Ecuador* (University of Texas Press, 2011), which won the National Women's Studies Association's Sara A. Whaley Book Prize. Her research focuses on issues of gender, bodies, and identity and has been published in journals such as *Gender & Society, Feminist Economics, Sociological Forum, Latino Studies, Women's Studies Quarterly*, and *Ethnography*. Her work has been featured in mainstream media outlets such as *The New York Times* and *The Huffington Post*, as well as international newspapers. She is currently conducting research on paid domestic work in Latin America and white-collar men and dress in corporate America.

Kimberly Kay Hoang is an assistant professor of Sociology and International Studies at Boston College and was formerly a postdoctoral fellow at Rice University. She received her PhD in Sociology with a Designated Emphasis in Women, Gender, and Sexuality from the University of California, Berkeley, in 2011. Her forthcoming book with the University of California Press studies niche markets in the global Vietnamese sex industry. She is interested in the links between changing political economies and intimacy, globalization and transnationalism, gender and migration, and changing urban spaces in Vietnam. Her work has appeared in *Sexualities* and the *Journal of Contemporary Ethnography*, as well as in media outlets like the BBC.

Susanne Hofmann is currently an honorary research fellow in the Department of Social Anthropology at the University of Manchester, UK. She holds a PhD in Latin American Cultural Studies. Her main research interests are bodies, commodification, corporeal entrepreneurship, transnational migration, neoliberalization, emotional labor, entrepreneurial subjectivities, and governmentality. She has published various articles on sex work in Mexico, including most recently in the *Women's Studies Quarterly*. She has a strong record of public engagement and community work beyond academia. She has been supporting queer migrants in Berlin and is an ally of sex worker unions demanding safe working conditions, respect, and workers' rights.

Afshan Jafar is an assistant professor of Sociology at Connecticut College. Her research and teaching interests are globalization, transnational women's movements, fundamentalist and nationalist movements, gender, and the body. Her first book, *Women's NGOs in Pakistan* (Palgrave Macmillan, 2011), uncovers the overwhelming challenges facing women's NGOs and examines the strategies used by them to ensure not just their survival but an acceptance of their messages by the larger public. Her research has been published in journals such as *Social Problems*,

Gender Issues, Critical Half and *Sexuality and Culture*. She is also a regular contributing editor for the University of Venus blog at InsideHigherEd.com and has also published career advice columns for InsideHigherEd.

Persis M. Karim is a professor in the Department of English and Comparative Literature at San Jose State University, where she teaches literature and creative writing. She has written numerous articles about Iranian American literature and is a poet and editor of several anthologies of Iranian American writing, including *Tremors: New Fiction by Iranian American Writers* (University of Arkansas Press, 2013); *Let Me Tell You Where I've Been: New Writing by Women of the Iranian Diaspora* (2006), and *A World Between: Poems, Short Stories and Essays by Iranian-Americans* (1999).

Fredrik Langeland is a cultural and masculinity researcher at the University of Stavanger in Norway. He is currently working on a PhD thesis on contemporary masculinities in Norwegian culture titled: "From Metro to Retro?" The project deals mainly with the representation of the metrosexual man in Norwegian media. He has published articles and book chapters on masculinities and sport, focusing on the representation of idealized bodies as symbols of the nation. He is a part of the research project *Being Together: Remaking Public Intimacies*, financed by the Norwegian Research Council. In 2011, he participated at the exhibition and book project *Together in This?* in which artists and researchers worked together in a postdisciplinary project on public intimacies.

Lucy Lowe is currently completing her PhD in Social Anthropology at the University of Edinburgh, Scotland. Her current research focuses on reproductive health, maternity, and onward migration among Somali refugees living in the Eastleigh area of Nairobi, commonly referred to as "Mogadishu Kidogo" [Little Mogadishu]. During her fieldwork she examined how displacement and desires for onward migration affect decisions relating to marriage, reproduction, and health, as well as exploring the tensions between pregnancy and childbirth as matters of both social and medical concern. She is also examining how such decisions are navigated within an urban context, with particular focus on perceptions of security, kinship, religion, and gender.

Michelle Newton-Francis is an assistant professor of Sociology at American University in Washington, DC. Her teaching and research interests lie at the intersection of popular culture, body and embodiment, and gender. She applies these interests to the study of erotic labor industries and cultural representations of female sexuality. Currently, she is studying the "Hooters Girl," including the ways in which the organization works to produce the gendered body, along with how the women inhabit their bodies as Hooters Girls and how they manage stigma related to their work.

Jaita Talukdar is an assistant professor in the Department of Sociology at Loyola University, New Orleans. Her research lies at the intersection of gender, culture, globalization, body, and health. She focuses on how modern individuals interpret and, in the process, modify social and cultural meanings about body and eating. Some of her research and book reviews have been published in journals such as *Women's Studies International Forum*, *Qualitative Sociology*, *Sociological Focus*, and the *American Journal of Sociology*. Her most recent research investigating socio-cultural factors associated with the growing popularity of biotechnological sciences of the body in contemporary Indian society has been accepted for publication in *The Handbook of Science, Technology, and Society* (to be published by Routledge in 2014). At present, she is studying how new upcoming gyms in the city of Kolkata, India are shaping perceptions and ideas of a fit and healthy body among urban Indians.

Salvador Vidal-Ortiz is an associate professor of Sociology at American University, in Washington, DC. He was a 2011 Fulbright scholar in Bogotá, Colombia, investigating the role sexual orientation and gender identity had in internal migration/displacement within the country. His research interests include racialization and US Latina/o communities, gender and sexuality, racialized sexualities, queer theory, social policy and HIV/AIDS, critical public health studies, the body, migration, and globalization, and religious-cultural practices like *santería*. He was the main convener, and the inaugural chair, of the section on Body and Embodiment of the American Sociological Association.

Jan Wickman received his doctorate in Sociology from Åbo Akademi University (Finland) in 2002. His research focuses on aspects of gender and sexuality in late and postmodernity, dealing with empirical themes such as collective identity building in the transgender community, sexualized representation of the male body in mass media, sports publicity, and, currently, aging among the LGBT population and queer activism. His key publications include *Transgender Politics: Construction and Deconstruction of Binary Gender in the Finnish Transgender Community* (2001), "Masculinity and Female Bodies," *NORA: The Nordic Journal of Women's Studies* 11 (2003): 40–54, *Queer* (together with Martin Berg, 2010) and "Queer Activism: What Might That Be?" Trikster—*Nor dic Queer Journal*, http://www.trikster.net/, Issue 4 (2011).

INDEX